Sharpening Everyday Mental/Thinking Skills
Through Mathematics Problem Solving and Beyond

Problem Solving in Mathematics and Beyond

Print ISSN: 2591-7234
Online ISSN: 2591-7242

Series Editor: Dr. Alfred S. Posamentier
Distinguished Lecturer
New York City College of Technology - City University of New York

There are countless applications that would be considered problem solving in mathematics and beyond. One could even argue that most of mathematics in one way or another involves solving problems. However, this series is intended to be of interest to the general audience with the sole purpose of demonstrating the power and beauty of mathematics through clever problem-solving experiences.

Each of the books will be aimed at the general audience, which implies that the writing level will be such that it will not engulfed in technical language — rather the language will be simple everyday language so that the focus can remain on the content and not be distracted by unnecessarily sophiscated language. Again, the primary purpose of this series is to approach the topic of mathematics problem-solving in a most appealing and attractive way in order to win more of the general public to appreciate his most important subject rather than to fear it. At the same time we expect that professionals in the scientific community will also find these books attractive, as they will provide many entertaining surprises for the unsuspecting reader.

Published

For the complete list of volumes in this series, please visit www.worldscientific.com/series/psmb

**Problem Solving in
Mathematics and Beyond** Volume **30**

Sharpening Everyday Mental/Thinking Skills Through Mathematics Problem Solving and Beyond

Alfred S. Posamentier
City University of New York, USA

Hans Humenberger
University of Vienna, Austria

 **World Scientific**

NEW JERSEY · LONDON · SINGAPORE · BEIJING · SHANGHAI · HONG KONG · TAIPEI · CHENNAI · TOKYO

Published by

World Scientific Publishing Co. Pte. Ltd.

5 Toh Tuck Link, Singapore 596224

USA office: 27 Warren Street, Suite 401-402, Hackensack, NJ 07601

UK office: 57 Shelton Street, Covent Garden, London WC2H 9HE

Library of Congress Cataloging-in-Publication Data

Names: Posamentier, Alfred S., author. | Humenberger, Hans, author.
Title: Sharpening everyday mental/thinking skills through mathematics problem
    solving and beyond / Alfred S. Posamentier (City University of New York, USA),
    Hans Humenberger (University of Vienna, Austria).
Description: New Jersey : World Scientific, [2024] | Series: Problem solving in
    mathematics and beyond, 2591-7234 ; volume 30 | Includes index.
Identifiers: LCCN 2023012622 | ISBN 9789811273940 (hardcover) |
    ISBN 9789811276392 (paperback) | ISBN 9789811273957 (ebook for institutions) |
    ISBN 9789811273964 (ebook for individuals)
Subjects: LCSH: Problem solving. | Problem solving--Methodology. | Mathematics.
Classification: LCC QA63 .P673 2024 | DDC 510--dc23/eng20230911
LC record available at https://lccn.loc.gov/2023012622

British Library Cataloguing-in-Publication Data
A catalogue record for this book is available from the British Library.

For any available supplementary material, please visit
https://www.worldscientific.com/worldscibooks/10.1142/13342#t=suppl

Desk Editors: Balasubramanian Shanmugam/Rosie Williamson

Typeset by Stallion Press
Email: enquiries@stallionpress.com

Printed in Singapore

About the Authors

Alfred S. Posamentier is currently a Distinguished Lecturer at the New York City College of Technology of the City University of New York. Prior to that, he was the Executive Director for Internationalization and Funded Programs at Long Island University, New York. This was preceded by 5 years as the Dean of the School of Education and Professor of Mathematics Education at Mercy University, New York. Before that, he was at the City College of the City University of New York for 40 years, at which he is now the Professor Emeritus of Mathematics Education and the Dean Emeritus of the School of Education. He is the author and co-author of more than 80 mathematics books for teachers, secondary and elementary school students, as well as the general readership. Dr. Posamentier is also a frequent commentator in newspapers and journals on topics related to education and mathematics.

After completing his B.A. degree in mathematics at Hunter College of the City University of New York, he took a position as a teacher of mathematics at Theodore Roosevelt High School (Bronx, New York), where he focused his attention on improving students' problem-solving skills and at the same time enriching their instruction far beyond what the traditional textbooks offered. During his 6-year tenure there, he also developed the school's first mathematics teams (both at the junior and senior levels). He is still involved in working with mathematics teachers and supervisors, nationally and internationally, to help them maximize their effectiveness.

Immediately upon joining the faculty of the City College of New York in 1970 (after having received his master's degree there in 1966), he began to develop in-service courses for secondary school mathematics teachers, including such special areas as recreational mathematics and problem solving in mathematics. As the Dean of the City College School of Education for 10 years, his scope of interest covered the full gamut of educational issues. During his tenure, he took the school from the bottom of the New York State rankings to the top with a perfect NCATE accreditation assessment in 2009. He also raised more than 12 million dollars from the private sector for innovative education programs. Dr. Posamentier repeated this successful transition at Mercy University, where he enabled it to become the only college to have received both NCATE and TEAC accreditation simultaneously.

In 1973, Dr. Posamentier received his Ph.D. from Fordham University (New York) in mathematics education and has since extended his reputation in mathematics education to Europe. He has been a visiting professor at several European universities in Austria, England, Germany, the Czech Republic, Turkey, and Poland. In 1990, he was the Fulbright Professor at the University of Vienna.

In 1989, he was awarded an Honorary Fellow position at South Bank University (London, England). In recognition of his outstanding teaching, the City College Alumni Association named him Educator of the Year in 1994 and 2009. New York City had the day, May 1, 1994, named in his honor by the President of the New York City Council. In 1994, he was also awarded the *Das Grosse Ehrenzeichen für Verdienste um die Republik Österreich* (Grand Medal of Honor from the Republic of Austria), and in 1999, upon approval of Parliament, the President of the Republic of Austria awarded him the title of *University Professor of Austria*. In 2003, he was awarded the title of *Ehrenbürgerschaft* (Honorary Fellow) of the Vienna University of Technology and, in 2004, was awarded the *Österreichisches Ehrenkreuz für Wissenschaft & Kunst 1.Klasse* (Austrian Cross of Honor for Arts and Science, First Class) from the President of the Republic of

Austria. In 2005, he was inducted into the Hunter College Alumni Hall of Fame, and in 2006 he was awarded the prestigious *Townsend Harris Medal* by the City College Alumni Association. He was inducted into the New York State Mathematics Educator's Hall of Fame in 2009, and in 2010 he was awarded the coveted *Christian-Peter-Beuth Prize* from the Technische Fachhochschule, Berlin. In 2017, Dr. Posamentier was awarded *Summa Cum Laude nemmine discrepante* by the Fundacion Sebastian, A.C., Mexico City, Mexico.

He has taken on numerous important leadership positions in mathematics education locally. He was a member of the New York State Education Commissioner's Blue Ribbon Panel on the Math-A Regents Exams, and the Commissioner's Mathematics Standards Committee, which redefined the Mathematics Standards for New York State, and he also served on the New York City schools' Chancellor's Math Advisory Panel.

Dr. Posamentier is still a leading commentator on educational issues and continues his long-time passion of seeking ways to make mathematics interesting to teachers, students, and the general public — as can be seen from some of his more recent books.

For more information and a list of his publications, see https://en.wikipedia.org/wiki/Alfred_S._Posamentier.

Hans Humenberger is currently a full Professor of Mathematics with special emphasis on mathematics education at the University of Vienna (Austria). He began his career studying mathematics and sports at the University of Vienna. In the 1990s he was a high school teacher at several Viennese high schools and also a graduate-assistant at the University of Natural Resources and Life Sciences, Institute for Mathematics, Vienna. In 1993 he earned his Ph.D. at the University of Vienna, and 1998 he received his habilitation in the field of mathematics education. Between 2000 and 2005 he was Assistant Professor of mathematics at the University of Dortmund (Germany). During this time, he also spent a half year as Interim Professor at the University of Duisburg-Essen (Germany). He returned to

Vienna in 2005 to accept his current professorship for mathematics at the University of Vienna. Since then, he has been the head of a working group "Didactics of Mathematics and School Mathematics" and also responsible for the educational training of preservice mathematics teachers for secondary school.

Of particular note is the forthcoming German title *Anschauliche Elementargeometrie,* written with his former colleague Berthold Schuppar, which is a geometry book intended largely for future teachers of mathematics.

He has been interested in mathematics problem solving for many years, which is reflected in several papers and corresponding seminars for preservice mathematics teachers, which he has held at several universities.

In 2007, he established an opportunity for secondary school students (grades 5–8) to attend seminars/workshops at the University of Vienna to explore interesting and challenging problems. This successful program is funded by the Vienna Department of Education and continues to the present day. Furthermore, he is particularly interested in making mathematics interesting for the general audience since this is important for the general perception of mathematics in our society.

Since 2007 he has been editor of an Austrian school textbook series in mathematics for grades 5–8, and in 2022 he was a member of a committee that established a new syllabus for mathematics at primary and secondary school levels in Austria.

His main fields of interest are mathematics as a process, applications of mathematics, problem solving, geometry, and stochastics.

More details and a complete list of his publications can be found at the following website: https://homepage.univie.ac.at/hans.humenberger/.

Contents

Contents

Introduction

This book is intended to share the usefulness, power and beauty of mathematics with the general public and beyond! Mathematics is a subject taught at all grades from kindergarten through high school, and yet it is the one subject that most adults are almost proud to admit not having been very good at and, therefore, tend to avoid it where they can. This, unfortunately, is sometimes passed onto children, which extends this negative legacy. This is regrettable as it indirectly tends to limit a person's reasoning ability, which is something that is trained throughout all of mathematics instruction — albeit oftentimes indirectly. One wonders why this continues to be the case decade after decade. At the elementary school level, teachers who reflect the general public are among those who are not very enamored with mathematics. Consequently, they do not exert themselves to make the subject interesting and exciting and are largely focused on "teaching to the test." We believe this is a huge disservice to the students. At the secondary school level, mathematics teachers have the opportunity to generate a new interest in mathematics, yet they too are concerned about student test performance and, therefore, rarely deviate from the standard curriculum. It is exactly that measured deviation that we believe is essential to demonstrate the beauty of mathematics, which in turn can generate or rekindle an interest among the students. There is a multitude of topics and skills in mathematics to which students are never exposed. In this book, we are attempting to demonstrate not only to the teachers but, in large measure, to the general public, the wonders that are hidden in mathematics and that can make it a very popular subject both in school and beyond.

Chapter 1 appropriately presents a wide variety of arithmetic calculations, which go far beyond the four basic operations: addition, subtraction, multiplication, and division. Here we investigate unusual and not well-known calculations and calculating shortcuts that not only give a genuine insight into arithmetic operations but also improve arithmetic skills both in writing and in thinking. This leads us nicely into Chapter 2, which presents a wide spectrum of numerical novelties that provides an extraordinary view of number relationships and numerical properties that can go a long way to impress the general readership about the hidden beauties in mathematics.

One of the key factors in mathematics is its ability to enable us to solve problems. Learning how to solve mathematical problems often has a favorable spin-off to solving everyday problems. For example, there are times when we deal with an everyday situation where we consider "the worst-case scenario" of the situation, which is analogous to solving a mathematical problem by considering extremes. Or we might consider the best path to take from point A to point B, where geometric relationships can be helpful. These are just some of the everyday problems that a broad background in mathematics problem solving can be supportive. Therefore, Chapter 3 of this book focuses on problem-solving strategies in mathematics, which are intended to be both entertaining and instructional. One of the most basic tools in mathematics is algebra, which allows us to understand the general case of certain ideas and concepts, and also to discover answers to complicated problems. Chapter 4 focuses on algebra and its somewhat unusual applications, which are intended to strengthen algebraic skills while at the same time demonstrating its power. As a side benefit, it contributes to a more favorable feeling towards the subject of mathematics in general.

Chapter 5 highlights the one aspect of mathematics that clearly demonstrates it beauty, which is geometry, perhaps the oldest branch of the field of mathematics, as it allows us to appreciate physical relationships that are sometimes counterintuitive and, thereby, can be highly motivating. The geometry that is offered here is beyond that which is taught at the secondary school level yet requires nothing more than some basics taught in school.

Finally, in Chapter 6 we provide a somewhat more challenging topic than offered in the previous chapters, as it explores a problem solving technique, which requires a bit of patience as it deals with problems in a stepwise procedure. That is, rather than to attack the problem with an immediate elegant solution, we present problems that can be very neatly solved by first considering various minor steps that will lead to a proper solution.

In its totality, this book is intended to demonstrate a variety of neglected aspects of mathematics that can regenerate interest in the subject for the general readership as well as for teachers who will, hopefully, provide a significant timeslot in their teaching each week to motivate students with some of the amazing aspects of mathematics we highlight. In short, we hope that this book will go a long way to popularize the field of mathematics by demonstrating its power and beauty beyond where most people believe is possible. So, join us now as we begin our journey through a myriad of surprising aspects of this most important subject: mathematics.

Chapter 1

Arithmetic Calculation Surprises

Although in today's culture arithmetic skills have been relegated to computers and calculators, there is still a great advantage to have insight into arithmetic calculations. They can not only be helpful and sometimes even swifter than a calculator, but also as a tool for appreciating mathematics and its logic. To multiply a two-digit number by 11 can be done more quickly mentally than with a calculator, if one knows that this product can be obtained by taking the sum of the digits of the two-digit number and placing it between the two digits as is the case with $53 \times 11 = 583$. Naturally, when the sum of the digits exceeds 9, slight adjustments must be made, such as carrying the 1 to the tens place. There are many astonishing arithmetic relationships that have been discovered over the past centuries. In this chapter, we will highlight merely a few with the hope that the reader will be motivated to seek out others. A note on symbols for multiplication in this book: There are various ways to indicate multiplication. The first one is the small $\times$; a second one is a centered dot; and a third one is simply using no symbol as with $2y$ or ab. In this book we will use all three versions as most convenient.

Clever Addition

One of the most popularly repeated stories from the history of mathematics is the tale of the famous mathematician Carl Friedrich Gauss who at age 10 was said to have mentally added the numbers from 1 to 100 in response to a busy work assignment given by the teacher.[1] Although it is a cute story and generally gets a very favorable reaction, it happens to provide us with a neat little formula for adding numbers in an arithmetic sequence, which is a list of numbers that has a common difference between consecutive numbers. For example, 2, 4, 6, 8, 10, ... is an arithmetic sequence as is each of the following:

$$5, 10, 15, 20, 25, 30, \ldots$$

$$7, 11, 15, 19, 23, 27, \ldots$$

Perhaps the simplest arithmetic sequence is the natural numbers: 1, 2, 3, 4, 5, ... What Gauss did to get the sum of the first 100 natural numbers without writing a single number was *not* to add the numbers in the order in which they appear but rather to add them in the following way:

the first plus the last,
the second plus the next-to-last,
the third plus the third from last,
and so on.

If we do this, we get the following:

$$1 + 100 = 101$$
$$2 + 99 = 101$$
$$3 + 98 = 101$$
$$4 + 97 = 101$$
$$\vdots$$
$$50 + 51 = 101$$

Note each pair has the same sum of 101. The sum of these 50 pairs of numbers is $50 \times 101 = 5{,}050$. From this example, we can derive a useful formula for adding numbers in an arithmetic sequence. We added the first and the last and multiplied this sum by one-half the number of members of

---

[1] According to E.T. Bell in his book, *Men of Mathematics* (New York: Simon & Schuster, 1937), the problem given to Gauss was of the sort: $81297 + 81495 + 81693 + \cdots + 100899$, where the common difference between consecutive terms was 198 and the number of terms was 100. Today's lore uses the numbers to be summed from 1 to 100, which makes the point just as well but in simpler form.

the sequence. We can generalize this and get a formula for an arithmetic series of n terms (with n an even number), where a is the first term and l is the last term (using Gauss' method) as follows: Sum $= \frac{n}{2}(a + l)$. This formula turns out to be correct also in the case that n is an odd number, where in this case $a + l$ is an even number. Here, we have an example of how simple it is to derive a very useful mathematical formula, one based on a very lovely pattern that wasn't completely obvious initially.

Surprising Addition

There is not much one can say beyond admiring the following results and appreciating the surprises that the addition of natural numbers offers:

$$1 + 2 = 3$$

$$4 + 5 + 6 = 7 + 8$$

$$9 + 10 + 11 + 12 = 13 + 14 + 15$$

$$16 + 17 + 18 + 19 + 20 = 21 + 22 + 23 + 24$$

$$25 + 26 + 27 + 28 + 29 + 30 = 31 + 32 + 33 + 34 + 35$$

etc.

Surprising Division

Some numbers lend themselves to a very beautiful symmetric division, which can be best demonstrated as follows:

$$121 = \frac{22 \times 22}{1 + 2 + 1} = \frac{484}{4} = 121$$

$$12321 = \frac{333 \times 333}{1 + 2 + 3 + 2 + 1} = \frac{110889}{9} = 12321$$

$$1234321 = \frac{4444 \times 4444}{1 + 2 + 3 + 4 + 3 + 2 + 1} = \frac{19749136}{16} = 1234321$$

$$123454321 = \frac{55555 \times 55555}{1 + 2 + 3 + 4 + 5 + 4 + 3 + 2 + 1} = \frac{3086358025}{25}$$
$$= 123454321$$

This pattern continues for four further steps until the digit 9 in the middle is reached.

Surprising Multiplication Results

When the number 076923 is multiplied by the numbers 1, 10, 9, 12, 3, and 4, quite a surprising pattern results; with the numbers staying in the same order, beginning with the multiplication by 10, the digit of the previous result far left is moved with each multiplication to the far right. Furthermore, if you inspect the vertical arrangements of the resulting numbers, you will find the same pattern.

$$076923 \times 1 = 076923$$

$$076923 \times 10 = 769230$$

$$076923 \times 9 = 692307$$

$$076923 \times 12 = 923076$$

$$076923 \times 3 = 230769$$

$$076923 \times 4 = 307692$$

An analogous pattern can be obtained by multiplying the number 76923 by the numbers 2, 7, 5, 11, 6, and 8 as follows:

$$76923 \times 2 = 153846$$

$$76923 \times 7 = 538461$$

$$76923 \times 5 = 384615$$

$$76923 \times 11 = 846153$$

$$76923 \times 6 = 461538$$

$$76923 \times 8 = 615384$$

Once again you will note the vertical arrangement of the digits of the numbers on the right side of the equal sign is analogous to the horizontal arrangements.

Using All Nine Digits to Make 100

The challenge here is to use all 9 digits 1, 2, 3, 4, 5, 6, 7, 8, and 9 to create an integer plus a fraction whose sum is 100. Here are 11 ways in which this can be done. An ambitious reader may wish to seek others.

$$3 + \frac{69258}{714}, \quad 81 + \frac{7524}{396}, \quad 81 + \frac{5643}{297}, \quad 82 + \frac{3546}{197}, \quad 91 + \frac{5823}{647},$$

$$91 + \frac{7524}{836}, \quad 91 + \frac{5742}{638}, \quad 94 + \frac{1578}{263}, \quad 96 + \frac{1428}{357}, \quad 96 + \frac{1752}{438},$$

$$96 + \frac{2148}{537}$$

Calculation Surprises

Consider the following addition and multiplication patterns. This is another example of the hidden beauty of mathematics.

$$\left(1 + \frac{1}{2}\right) \times 3 = 4\frac{1}{2} = 1 + \frac{1}{2} + 3$$

$$\left(1 + \frac{1}{3}\right) \times 4 = 5\frac{1}{3} = 1 + \frac{1}{3} + 4$$

$$\left(1 + \frac{1}{4}\right) \times 5 = 6\frac{1}{4} = 1 + \frac{1}{4} + 5$$

$$\left(1 + \frac{1}{5}\right) \times 6 = 7\frac{1}{5} = 1 + \frac{1}{5} + 6$$

$$\vdots$$

$$\left(1 + \frac{1}{n}\right) \times (n + 1) = (n + 2)\frac{1}{n} = 1 + \frac{1}{n} + (n + 1)$$

Consider the following multiplication and subtraction pattern:

$$1 \times \frac{1}{2} = \frac{1}{2} = 1 - \frac{1}{2}$$

$$2 \times \frac{2}{3} = 1\frac{1}{3} = 2 - \frac{2}{3}$$

$$3 \times \frac{3}{4} = 2\frac{1}{4} = 3 - \frac{3}{4}$$

$$4 \times \frac{4}{5} = 3\frac{1}{5} = 4 - \frac{4}{5}$$

$$5 \times \frac{5}{6} = 4\frac{1}{6} = 5 - \frac{5}{6}$$

$$\vdots$$

$$n \times \frac{n}{n+1} = (n-1) + \frac{1}{n+1} = n - \frac{n}{n+1}$$

Consider the following symmetry involving division and addition:

$$1\frac{1}{3} \div \frac{2}{3} = 2 = 1\frac{1}{3} + \frac{2}{3}$$

$$2\frac{1}{4} \div \frac{3}{4} = 3 = 2\frac{1}{4} + \frac{3}{4}$$

$$3\frac{1}{5} \div \frac{4}{5} = 4 = 3\frac{1}{5} + \frac{4}{5}$$

$$4\frac{1}{6} \div \frac{5}{6} = 5 = 4\frac{1}{6} + \frac{5}{6}$$

$$\vdots$$

$$\left(n + \frac{1}{n+2}\right) \div \frac{n+1}{n+2} = n+1 = \left(n + \frac{1}{n+2}\right) + \frac{n+1}{n+2}$$

Lastly, consider a symmetry between division and subtraction:

$$4\frac{1}{2} \div 3 = 1\frac{1}{2} = 4\frac{1}{2} - 3$$

$$5\frac{1}{3} \div 4 = 1\frac{1}{3} = 5\frac{1}{3} - 4$$

$$6\frac{1}{4} \div 5 = 1\frac{1}{4} = 6\frac{1}{4} - 5$$

$$7\frac{1}{5} \div 6 = 1\frac{1}{5} = 7\frac{1}{5} - 6$$

$$\vdots$$

$$n + \left(\frac{1}{n-2}\right) \div (n-1) = 1 + \left(\frac{1}{n-2}\right) = n + \left(\frac{1}{n-2}\right) - (n-1)$$

Squaring a Multiple of 5

Consider the challenge of squaring the number 85 mentally. Here is a way that you can perform this calculation rather easily. All we need to do, following this curious technique, is to multiply the tens digit, which is 8, by $8 + 1 = 9$, then multiply that by 100 and add 25, as shown here: $85^2 = (8 \times 9 \times 100) + 25 = 7225$. With a little practice, this can go quite quickly and provide another useful arithmetic technique. To see why this works, let's consider a general case of a two-digit number $(10a + 5)$, which is a multiple of 5. When we square this number, we get $(10a + 5)^2 = 100a^2 + 100a + 25 = a(a+1)(100) + 25$.

Squaring Numbers from 1 to 125

This technique requires us to partition the numbers from 1 to 125 into three groups:

(1) The squares of the first 25 numbers 1–25 are most likely known by many people without calculation.

The squares of the numbers 25–70 are symmetric about the number 50, as 25–**50**–75, and their squares can be calculated easily with the following technique:

Add to (subtract from) 25 the distance of the number to 50 and append on the right its square as a two-digit number (however, if this square is a three-digit number, then carry its hundreds digit to the last digit of the first operation).

This can be easily shown with the following examples:

$57^2 : 25 + 7 = 32$; $7^2 = 49$ 32 |49, which is the number 3,249

$36^2 : 25 - 14 = 11$; $14^2 = 196$, then $11 + 1 \mid 96$ or 12 |96, which is the number 1,296

This always works because $(50 \pm a)^2 = 2500 \pm 100a + a^2 = (25 \pm a) \cdot 100 + a^2$.

If one allows a to also be negative, then one does not need to distinguish between $50 \pm a$, one can just write $(50 + a)^2 = 2500 + 100a + a^2 = (25 + a) \cdot 100 + a^2$.

(2) The squares of the numbers between 75 and 125 (which are symmetric about 100 as we can see with 75–**100**–125) can be calculated easily using the following technique:

Add to (subtract from) the number its distance to 100 and then append on the right the squared distance as a two-digit number.

Consider the following examples:

$106^2 : 106 + 6 = 112$; $6^2 = 36$ 112 | 36, which is the number 11,236

$88^2 : 88 - 12 = 76$; $12^2 = 144$ then $76 + 1|44$ or 77 | 44, which is the number 7,744

The reason that this works can be shown as follows: $(100 \pm a)^2 = 100^2 \pm 2(100a) + a^2 = (100 \pm 2a)(100) + a^2 = [(100 \pm a) \pm a](100) + a^2$

If one allows a to also be negative, then one does not need to distinguish between $100 \pm a$.

The Unusual Number 9

The first occurrence in Western Europe of the Hindu–Arabic numerals we use today was in the book *Liber abaci* written in 1202 by Leonardo of Pisa (commonly known as Fibonacci).

He begins the book with an introduction that reads:

these are the nine figures of the Indians 9, 8, 7, 6, 5, 4, 3, 2, 1. With these nine figures, and with the symbol, 0, which in Arabic is called zephirum, any number can be written, as will be demonstrated . . .

With this book, the use of these numerals was first popularized in Europe. Before that, Roman numerals were used, which were clearly much more cumbersome to use for calculation. Fascinated as he was by the arithmetic calculations used in the Islamic world, Fibonacci, in his book, first introduced the system of "casting out nines"[2] as a check for arithmetic. It comes in handy even today. However, the nice thing about it is that it again demonstrates a hidden magic in ordinary arithmetic.

Before we discuss this arithmetic-checking procedure, we will consider how the remainder of a division by 9 compares to removing nines from the digit sum of the number. When 8,768 is divided by 9, the quotient is 974 with a remainder of 2. This remainder can also be obtained by "casting out nines" from the digit sum of the number 8,768: We will find the sum of the digits and, if the sum is more than a single digit, we shall repeat the procedure. In the case of our given number, 8,768, the digit sum is 29 ($8 + 7 + 6 + 8 = 29$) so we will repeat the process. Again, the casting-out-nines procedure is used to get $2 + 9 = 11$, and again $1 + 1 = 2$, which was, in fact, the remainder when we divided 8,768 by 9.

We can use casting out nines to check if a multiplication problem is correct. Suppose we would like to multiply 734×879. The answer we get is 645,186. Now casting out nines by taking digit sums, we get the following:

For 734: $7 + 3 + 4 = 14$;   then $1 + 4 = 5$
For 879: $8 + 7 + 9 = 24$;   then $2 + 4 = 6$
For the product of these two
numbers, 645,186: $6 + 4 + 5 + 1 + 8 + 6 = 30$

Since $5 \times 6 = 30$, which yields 3 (casting out nines: $3 + 0 = 3$), is the same as for the sum of the digits of the product, the answer could be correct. For practice, we will do another casting-out-nines "check" for the following multiplication: $56,589 \times 983,678 = 55,665,354,342$.

For 56,589 $5 + 6 + 5 + 8 + 9 = 33$; $3 + 3 = 6$
For 983,678 $9 + 8 + 3 + 6 + 7 + 8 = 41$; $4 + 1 = 5$

---

[2]"Casting out nines" refers to an arithmetic check that tells you if your answer is possibly correct. The process requires taking bundles of nines away from the sum of the digits.

The product of the remainders: $5 \times 6 = 30 \rightarrow 3 + 0 = 3$, compares favorably to the digit-sum remainder as we see here:

$55,665,354,342: 5 + 5 + 6 + 6 + 5 + 3 + 5 + 4 + 3 + 4 + 2 = 48;$

$4 + 8 = 12; \ 1 + 2 = 3$

The same procedure can be used to check the likelihood of a correct sum, difference, or quotient, simply by taking the sum (difference or quotient) and casting out nines, taking the sum (difference or quotient) of these "remainders," and comparing it with the remainder of the sum (difference or quotient). If these digit-sum remainders are not equal the answer is surely wrong. If these digit-sum remainders are equal, the answer has a good chance to be correct.

More Novelties with the Number 9

The number 9 has another unusual feature that enables us to use a surprising multiplication algorithm. Although it is somewhat complicated, it is nevertheless fascinating to see how it functions and perhaps try to determine why this actually works. This procedure is intended for multiplying a number of two digits or more by 9. It is perhaps best to discuss the procedure in the context of an actual example: Consider multiplying $76,354 \times 9$. Now follow the steps shown in Figure 1.1.

Step 1	Subtract the units digit of the multiplicand from 10	$10 - 4 = \mathbf{6}$
Step 2	Subtract each of the remaining digits (beginning with the tens digit) from 9 and add this result to the previous digit in the multiplicand (for any two digit sums, carry the tens digit to the next sum)	$9 - 5 = 4, \ 4 + 4 = \mathbf{8}$ $9 - 3 = 6, \ 6 + 5 = 11, \ \mathbf{1}$ $9 - 6 = 3, \ 3 + 3 = 6, \ 6 + 1 = \mathbf{7}$ $9 - 7 = 2, \ 2 + 6 = \mathbf{8}$
Step 3	Subtract 1 from the left-most digit of the multiplicand	$7 - 1 = \mathbf{6}$
Step 4	List the results in reverse order to get the desired product	$\mathbf{687,186}$

Figure 1.1

Although it is a bit cumbersome, especially when compared to a calculator, this algorithm provides some insights into number theory.

Non-Repeating Eight-Digit Numbers and the Number 9

There are many eight-digit numbers with no digits repeated that when multiplied by 9 yield nine-digit numbers which have some digits repeated, such as $56,387,412 \times 9 = 507,486,708$. However, there are some instances in which a non-repeating-digit eight-digit number multiplied by 9 results in a nine-digit number with no repeating digits such as

$$81274365 \times 9 = 731469285$$

$$76125484 \times 9 = 685129347$$

$$72645831 \times 9 = 653812479$$

$$58132764 \times 9 = 523194876$$

Curiously enough, when these four products of nine digits are multiplied by 2, the result is once again a 10-digit number with no repeating digits as we can see with the following:

$$731469285 \times 2 = 1462938570$$

$$685129347 \times 2 = 1370258694$$

$$653812479 \times 2 = 1307624958$$

$$523194876 \times 2 = 1046389752$$

Once again, the number 9 plays an important role in our arithmetic wonders!

The Magic of the Number 9

The only peculiarity of the number 12,345,679 is that it has the digits in sequence and is missing the number 8. If we multiply this number by products of 9 with any of the numbers 1, 2, 3, 4, 5, 6, 7, 8, and 9, we get surprising results:

$$9 \times 1 \times 12,345,679 = 111,111,111$$

$$9 \times 2 \times 12,345,679 = 222,222,222$$

$$9 \times 3 \times 12,345,679 = 333,333,333$$

$$9 \times 4 \times 12,345,679 = 444,444,444$$

$$9 \times 5 \times 12,345,679 = 555,555,555$$

$$9 \times 6 \times 12,345,679 = 666,666,666$$

$$9 \times 7 \times 12,345,679 = 777,777,777$$

$$9 \times 8 \times 12,345,679 = 888,888,888$$

$$9 \times 9 \times 12,345,679 = 999,999,999$$

This unusual feature can be easily explained, since $n \times 9 = n(10 - 1)$. Now multiplying the number $n(10 - 1)$ by 12,345,679 gives us $n(123,456,790 - 12,345,679) = n(111,111,111)$.

A Secret about the Number 9

There is a curiosity when the number 15 is placed between the digits of the number $4^2 = 16$.

1**15**6 $= 34^2$.

When we place the number 15 in the middle of the number 1156, we get 11**15**56 $= 334^2$.

When we place the number 15 in the middle of the number 111556, we get 111**15**556 $= 3334^2$.

When we place the number 15 in the middle of the number 11115556, we get 1111**15**5556 $= 33334^2$. This pattern continues. The problem is to show why this continues to work further on.

Solution: We can understand this pattern by considering any one of the numbers in our pattern, say, 11115556, which when multiplied by 9 is equal to 100040004 $= 10002^2$. This can be shown to hold true by analyzing the product as follows: 11115556 $\times (10 - 1) = 111155560 - 11115556 = 100040004$. At this point, we should note that numbers of the form "1000...4000...4" are always square numbers. Remember this was generated by the number 9.

There is one other number, 48, that can analogously generate square numbers. This time we begin with the number 49 and place the number

48 between the center digits so that the next number will be $4\underline{48}9 = 67^2$. And then $444\underline{8}89 = 667^2$, and $44\underline{448}889 = 6667^2$, and so on. Once again, it is the number 9 that generates this pattern as follows: $444889 \times 9 = 4004001 - 2001^2$.

The Technique for Determining When a Number is Divisible by 3 or 9

There are moments in everyday life situations where it can be useful to know if a number can be divisible by 3 or by 9, especially if it can be done instantly "in your head." For example, if you are in a restaurant and receive a check of $71.23, and you want to add a tip but you would want the end result to be able to be split evenly among three people. Wouldn't it be nice if there were some mental arithmetic technique for determining this? Well, here comes mathematics to the rescue. We are going to provide you with a simple technique to determine if a number is divisible by 3 and (as an extra bonus) also divisible by 9.

The rule, simply stated, is as follows:

If, and only if, the sum of the digits of a number is divisible by 3 (or 9), the original number is divisible by 3 (or 9).

As before, perhaps an example of this technique would best firm up an understanding of how it works. Consider the number 296,357. Let's test it for divisibility by 3 (or 9). The sum of the digits is $2+9+6+3+5+7 = 32$, which is not divisible by 3 or 9. Therefore, the original number is neither divisible by 3 nor 9.

Let's now assume a group of three people is given a restaurant check of $71.23 and would like to give an approximate 20% tip on this check. They decide to add $14 to the bill, which would make the total $85.23. To check for divisibility by 3, since they would like to divide the check equally among the three persons. The three participants would have them add the digits to get $8 + 5 + 2 + 3 = 18$, which is divisible by 3 and, therefore, would allow them to equally divide the check among the three customers. If the number reached from the sum of the digits is not easily identifiable as a multiple of 3, then continue to add the digits of that number until you reach a number that you can easily recognize as a multiple of 3. In this case, it should be noted that the final result, 18, is also divisible by 9, which implies that the original number was divisible by 9 as well.

Here is a brief explanation about why this rule works. Consider the number *ab,cde*, whose value can be expressed as (where 9*M* refers to a multiple of 9)

$$N = 10^4a + 10^3b + 10^2c + 10d + e$$
$$= (9+1)^4a + (9+1)^3b + (9+1)^2c + (9+1)d + e$$
$$= [9M + (1)^4]a + [9M + (1)^3]b + [9M + (1)^2]c + [9 + (1)]d + e$$
$$= 9M[a+b+c+d] + a+b+c+d+e$$

which implies that divisibility by nine of the number N depends on the divisibility of $a+b+c+d+e$, which is the sum of the digits.

A few examples to firm up this technique are as follows:

Is the number 457,875 divisible by 3 or 9? The sum of the digits is $4+5+7+8+7+5 = 36$, *which is divisible by 9 (and then, of course, by 3 as well), so the number 457,875 is divisible by 3 and by 9.*

Is the number 27,987 divisible by 3 or 9? The sum of the digits is $2 + 7+9+8+7 = 33$, *which is divisible by 3 but not by 9. Therefore, the number 27,987 is divisible by 3 but not by 9.*

Now that you are an expert at determining if a number is divisible by 3 or 9, we can go back to the original question about the divisibility of a restaurant bill of $71.22 + 14.00 = $85.22. Can it be divided into three equal parts? Because $8 + 5 + 2 + 2 = 17$, and 17 is not divisible by three, therefore $85.22 is not divisible by 3, and they will have to modify the tip to get everyone to pay the same amount.

Multiplying by 11 Mentally

Multiplying by 11 can be significantly simpler than using a calculator, and yet, this technique is hardly ever presented in school. What a shame! When multiplying by 11, the technique for doing it is as fast as you can write the number you are multiplying by 11. For example, suppose you would like to multiply 23 × 11. All you need to do is add the digits, $2 + 3 = 5$, and place that digit 5 between the two digits in the number 23 to get 253. What could be easier! We can state this technique as follows:

To multiply a two-digit number by 11, just add the two digits and place this sum between the two digits. If the sum of the digits is a two-digit number, then carry the tens digit of the sum to be added to the former tens digit.

Just for practice, we will apply this technique to multiply 45 by 11. According to the rule, add 4 and 5 and place it between the 4 and 5 to get 495. This technique can get a bit more difficult when the sum of the two digits you need to add results in a two-digit number. We no longer have a single digit to place between the two original digits. So, if the sum of the two digits is greater than 9, then we would place the units digit of that sum between the two digits of the number being multiplied by 11 and "carry" the tens digit to be added to the hundreds digit of the multiplicand. Let's try it with 78 × 11. We find that the sum of the two digits is $7 + 8 = 15$. Therefore, we place the 5 between the 7 and 8 and add the 1 to the 7 to get [7 + 1][5][8] or 858.

You may legitimately ask if the rule also holds when 11 is multiplied by a number of more than two digits. Consider a larger number such as 12,345 and use this technique to multiply it by 11.

Here, we begin at the right-side digit and add every pair of digits going from right to left.

$$1[1 + 2][2 + 3][3 + 4][4 + 5]5 = 135{,}795.$$

Suppose we now combine the skill we have garnered through this technique of multiplying by 11 and apply it to a number, which requires a more complicated version, where the sum of adjacent digits exceeds 9. Remember, if the sum of two digits is greater than 9, then use the procedure described earlier: Place the units digit of this two-digit sum appropriately and carry the tens digit to the next place. To become an expert at this procedure, we will do one of these more complicated versions here. Let us consider multiplying 56,789 by 11. This may be a little bit tedious, and perhaps somewhat less realistic for common use, but we will show it here merely to demonstrate the extension of this multiplication technique. Follow along as we do this step by step.

5[5 + 6][6 + 7][7 + 8][8 + 9]9	Add each pair of digits between the end digits
5[5 + 6][6 + 7][7 + 8][17]9	Add 8 + 9 = 17
5[5 + 6][6 + 7][7 + 8 + 1][7]9	Carry the 1 (from the 17) to the next sum
5[5 + 6][6 + 7][16][7]9	Add 7 + 8 + 1 = 16
5[5 + 6][6 + 7 + 1][6][7]9	Carry the 1 (from the 16) to the next sum
5[5 + 6][14][6][7]9	Add 6 + 7 + 1 = 14
5[5 + 6 + 1][4][6][7]9	Carry the 1 (from the 14) to the next sum
5[12][4][6][7]9	Add 5 + 6 + 1 = 12
5 + 1[2][4][6][7]9	Add 5 + 1 = 6 to get the answer 624,679

This technique for multiplying by 11 could actually be helpful when least anticipated.

A Technique to Determine When a Number Is Divisible by 11

One can also be fortified with another novelty of the number 11, which owes its special status to being 1 greater than the base 10 of our number system. Let's look at the reverse of the previous arithmetic technique, that is, division by 11 rather than multiplication by 11. At the oddest times, the need can come up to determine if a number is divisible by 11. If you have a calculator at hand, the problem is easily solved. But that is not always the case. Besides, there is such a clever "rule" for testing for divisibility by 11 that it is worth knowing it even just for its charm.

The rule is quite simple:

If, and only if, the difference of the sums of the alternate digits is divisible by 11, the original number is also divisible by 11.

This may sound more complicated than it actually is. Let us take this rule a piece at a time. The sums of the alternate digits mean that you begin at one end of the number taking the first, third, fifth, etc., digits and adding them. Then add the remaining (even placed) digits. Subtract the two sums and inspect for divisibility by 11. If the resulting number is divisible by 11, then the original number was divisible by 11. And the reverse is also true. That is, if the number reached by subtracting the two sums is *not* divisible by 11, then the original number was also *not* divisible by 11.

This is probably best demonstrated by an example. Suppose we test the number 768,614 for divisibility by 11. Sums of the alternate digits are $7 + 8 + 1 = 16$ and $6 + 6 + 4 = 16$. The difference of these two sums $16 - 16 = 0$, which is divisible by 11. (Remember $\frac{0}{11} = 0$.) Therefore, we can conclude that 768,614 is divisible by 11.

Another example might be helpful to firm up your understanding of this procedure. To determine if 918,082 is divisible by 11, we need to find the sums of the alternate digits:

$$9 + 8 + 8 = 25 \quad \text{and} \quad 1 + 0 + 2 = 3$$

Their difference is $25 - 3 = 22$, which is divisible by 11, and so the number 918,082 is divisible by 11. Practice will make this technique quite useful.

Here is a brief explanation to justify this rule. Consider the number ab,cde, whose value can be expressed as follows (here $11M$ represents a multiply of 11):

$$
\begin{aligned}
N &= 10^4 a + 10^3 b + 10^2 c + 10d + e \\
&= (11 - 1)^4 a + (11 - 1)^3 b + (11 - 1)^2 c + (11 - 1)d + e \\
&= [11M + (-1)^4]a + [11M + (-1)^3]b + \\
&\quad [11M + (-1)^2]c + [11 + (-1)]d + e \\
&= 11M[a + b + c + d] + a - b + c - d + e,
\end{aligned}
$$

which implies that since the first part of this value of N, namely, $11M[a + b + c + d]$, is already a multiple of 11, the divisibility by 11 of the number N depends on the divisibility of the remaining part, which is $a - b + c - d + e = (a + c + e) - (b + d)$ and which is actually the difference of the sums of the alternate digits.

Divisibility by Prime Factors of 10 and Their Powers

Most people can determine if a number is divisible by 2 or by 5, simply by looking at the last digit (i.e., the units digit) of the number. That is, if, and only if, by looking the last digit is an even number (such as 2, 4, 6, 8, and 0), then the number will be divisible by 2. Incidentally, you can extend this simple well-known technique to one where you can determine divisibility by higher powers of 2, such as the divisibility by 4, 8, 16, etc. Once again, we look at the end of the number being inspected. If, and only if, the number formed by the last two digits is divisible by 4, then the entire number itself is divisible by 4. Also, if, and only if, the number formed by the last three digits is divisible by 8, then the number itself is divisible by 8. And so, to test for divisibility by 16, we would focus on the last four digits of the number, and so on for other higher powers of 2.

In an analogous fashion, we can develop a technique for determining divisibility by powers of 5. It is common knowledge that if the last digit of the number being inspected for divisibility is either 0 or 5, then the number itself will be divisible by 5. A number is divisible by 25 if, and only if, the last two digits of the number, taken together as a number, is divisible by 25. This is analogous to the technique we used to determine divisibility by powers of 2. Have you guessed what the relationship here is between powers of 2 and 5? Yes, they are both factors of 10, the base of our decimal number system.

Divisibility by Other Prime Numbers

Divisibility by 7

With the proliferation of the calculator, there is no longer a crying need to be able to detect by which numbers a given number is divisible. Yet, for a better appreciation of mathematics, divisibility rules provide an interesting "window" into the nature of numbers and their properties. For this reason (among others), the topic of divisibility still finds a place on the mathematics-learning spectrum and can provide some further problem-solving strength. This is especially true for the technique for divisibility by 7. (By the way, the technique to determine divisibility by 6 is simply to apply the technique for divisibility by 2 and by 3 — both must hold true for a number to be divisible by 6.)

We begin by considering the method to determine if a given number is divisible by 7 and then, as we inspect this technique, we can see how it can be generalized for other prime numbers.

The technique for divisibility by 7 is as follows:

Delete the last digit from the given number and then subtract twice this deleted digit from the remaining number. If, and only if, the result is divisible by 7, the original number is divisible by 7. This process may be repeated until the result can be determined by simple inspection for divisibility by 7.

For example, suppose we want to test the number 876,547 for divisibility by 7. Begin with 876,547 and delete its units digit, 7, and subtract its double, 14, from the remaining number: $87,654 - 14 = 87,640$. Since we cannot yet visually inspect the resulting number for divisibility by 7, we continue the process with the resulting number 87,640 and delete its units digit, 0, and subtract its double, still 0, from the remaining number; we get $8,764 - 0 = 8,764$. This did not bring us any closer to visually being able to check for divisibility by 7; therefore, we continue the process with the resulting number 8,764 and delete its units digit, 4, and subtract its double, 8, from the remaining number; we get $876 - 8 = 868$. Since we still cannot visually inspect the resulting number, 868, for divisibility by 7, we continue the process with the resulting number 868 and delete its units digit, 8, and subtract its double, 16, from the remaining number we get $86 - 16 = 70$, which we can easily see is divisible by 7. Therefore, the original number 876,547 is also divisible by 7. Before continuing with our exploration of divisibility by prime numbers, it would be wise to practice this technique with a few randomly selected numbers and then check your results with a calculator.

Why does this rather strange procedure work? Being able to answer this question is the wonderful thing about mathematics. It doesn't do things that, for the most part, we cannot justify. This will all make sense after you see what is happening with this procedure. To justify the technique of determining divisibility by 7, consider the various possible terminal digits (that you are "dropping") and the corresponding subtraction that is actually being done by dropping the last digit. In the chart (Figure 1.2), you will see that by dropping the terminal digit and doubling it, the number being subtracted gives us in each case a multiple of 7. That is, we have taken "bundles of 7" away from the original number. Therefore, if, and only if, the the remaining number is divisible by 7, then so is the original number divisible by 7 because you have separated the original number into two parts, each of which is divisible by 7, and, therefore, this must hold true for the entire number.

Terminal digit	Number subtracted from original
1	$20 + 1 = \ 21 = \ 3 \times 7$
2	$40 + 2 = \ 42 = \ 6 \times 7$
3	$60 + 3 = \ 63 = \ 9 \times 7$
4	$80 + 4 = \ 84 = 12 \times 7$
5	$100 + 5 = 105 = 15 \times 7$
6	$120 + 6 = 126 = 18 \times 7$
7	$140 + 7 = 147 = 21 \times 7$
8	$160 + 8 = 168 = 24 \times 7$
9	$180 + 9 = 189 = 27 \times 7$

Figure 1.2

Divisibility by 13

The technique for divisibility by 13 is very similar to that for divisibility by 7 except that the 7 is replaced by 13, and instead of subtracting twice the deleted digit, we subtract nine times the deleted digit each time. It thus reads as follows:

Delete the last digit from the given number and then subtract nine times this deleted digit from the remaining number. If, and only if, the result is divisible by 13, the original number is divisible by 13.

Let's try our technique to check for divisibility by 13 for the number 5,616. We begin by deleting its units digit, 6, and this time instead of doubling the number to subtract, we subtract nine times the number, in this case, $9 \times 6 = 54$, from the remaining number $561 - 54 = 507$. Since we still cannot visually inspect the resulting number for divisibility by 13, we continue the process with the resulting number 507 and delete its units digit and subtract 9 times this digit ($9 \times 7 = 63$) from the remaining number

$50 - 63 = -13$, which is divisible by 13, and, therefore, the original number is also divisible by 13.

Now we might want to see how we determined the "multiplier" 9 in our technique. We sought the smallest multiple of 13 that ends in 1. That was 91, where the tens digit is 9 times the units digit. Once again, consider the various possible terminal digits and the corresponding subtractions in Figure 1.3.

Terminal digit	Number subtracted from original
1	$90 + 1 = 91 = 7 \times 13$
2	$180 + 2 = 182 = 14 \times 13$
3	$270 + 3 = 273 = 21 \times 13$
4	$360 + 4 = 364 = 28 \times 13$
5	$450 + 5 = 455 = 35 \times 13$
6	$540 + 6 = 546 = 42 \times 13$
7	$630 + 7 = 637 = 49 \times 13$
8	$720 + 8 = 728 = 56 \times 13$
9	$810 + 9 = 819 = 63 \times 13$

Figure 1.3

In each case, a multiple of 13 is being subtracted one or more times from the original number. Hence, if, and only if, the remaining number is divisible by 13, then the original number is divisible by 13.

Divisibility by 17

The technique for divisibility by 17 is as follows:

Delete the units digit and subtract five times the deleted digit each time from the remaining number until you reach a number small enough to visually determine its divisibility by 17.

We justify the technique for divisibility by 17 as we did the techniques for 7 and 13. Each step of the process requires us to subtract a "bunch of 17s" from the original number until we reduce the number to a manageable size by which we can make a visual inspection of divisibility by 17. This time we can see that the multiplier is 5, since we will be deducting multiples of 17, such as 51, 102, 153, and so on, from the original number.

Divisibility by larger prime numbers

The patterns developed in the preceding three divisibility techniques (for 7, 13, and 17) should enable you to produce analogous techniques for testing divisibility by larger primes. Figure 1.4 presents the "multipliers" of the deleted digits for various primes.

To test divisibility by	7	11	13	17	19	23	29	31	37	41	43	47
Multiplier	2	1	9	5	17	16	26	3	11	4	30	14

Figure 1.4

You may want to extend this chart as it will increase your perception of mathematics, while at the same time extending your toolkit of problem-solving techniques. You may also want to extend your knowledge of divisibility rules to include composite (i.e., non-prime) numbers. Why the following rule refers to relatively prime factors and not just any factors is something that will sharpen your understanding of number properties. Perhaps the easiest response to this question is that relatively prime factors have independent divisibility rules, whereas other factors may not.

The following technique is for divisibility by composite numbers:

A given number is divisible by a composite number, if, and only if, it is divisible by each of its relatively prime factors. (Two or more numbers are relatively prime if they have no common factors other than 1.)

Figure 1.5 offers illustrations of this rule.

To be divisible by	6	10	12	15	18	21	24	26	28
The number must be divisible by	2, 3	2, 5	3, 4	3, 5	2, 9	3, 7	3, 8	2, 13	4, 7

Figure 1.5

To this point, you have had an opportunity to broaden your arithmetic problem-solving skills that give you a deeper insight into mathematics as well as an appreciation for its wonders. There are times when a calculator is not at hand, and we need to do mental arithmetic. Thus, we continue on with arithmetic problem-solving skills.

Amazing Sum of Squares

Consider any eight consecutive numbers, such as 6, 7, 8, 9, 10, 11, 12, and 13. Quite unexpectedly, the squares of these eight numbers can be organized into equal parts such as $6^2 + 9^2 + 11^2 + 12^2 = 382 = 7^2 + 8^2 + 10^2 + 13^2$. Analogously, we can show that for the eight consecutive numbers 7, 8, 9, 10, 11, 12, 13, and 14 we have the equal sums as follows: $7^2 + 10^2 + 12^2 + 13^2 = 462 = 8^2 + 9^2 + 11^2 + 14^2$. Why does this happen?

Solution: We can best explain this amazing phenomenon using simple algebra. Based on the pattern above, where we let n represent the first number, $(n+1)$ the second number, ... $(n+7)$ the eighth number, we have established the following pattern:

$$n^2 + (n + 3)^2 + (n + 5)^2 + (n + 6)^2 = 4n^2 + 28n + 70$$

and the remaining numbers used:

$$(n + 1)^2 + (n + 2)^2 + (n + 4)^2 + (n + 7)^2 = 4n^2 + 28n + 70.$$

This proves that for any eight consecutive numbers, this pattern will hold true.

A Squaring Curiosity

Consider the amazing symmetry by squaring the sum of two numbers where the result is a number formed by the cumulation of the digits. For example, $(5288 + 1984)^2 = 52,881,984$. Are there other numbers that share this curious property?

Solution: The answer to the question is clearly yes. One such example would be $(6048 + 1729)^2 = 60,481,729$. What we seek are numbers where the following would be true: $(a + b)^2 = ab$, or $(ab + cd)^2 = abcd$, where ab is a two-digit number and $abcd$ is a four-digit number. In the first case, we can have $(0 + 1)^2 = 01$ or $(8 + 1)^2 = 81$. For pairs of two-digit numbers, some examples are as follows:

$$(20 + 25)^2 = 2,025$$

$$(30 + 25)^2 = 3,025$$

$$(98 + 01)^2 = 9,801$$

This last number when divided by 9 yields the reversal 1,089.

In the meantime, we can also consider pairs of three-digit numbers that share this unusual characteristic.

$$(088 + 209)^2 = 088,209$$

$$(494 + 209)^2 = 494,209$$

$$(998 + 001)^2 = 998,001$$

There are even four-digit numbers that exhibit this amazing symmetry such as

$$(9998 + 0001)^2 = 99,980,001$$

The Russian Peasant's Method of Multiplication

It is said that the Russian peasants used a rather strange, perhaps even primitive, method to multiply two numbers. It is actually quite simple, yet somewhat cumbersome. Consider the problem of finding the product of 43×92. We begin by setting up a chart of two columns with the two members of the product in the first row, as shown in Figure 1.6. One column will be formed by doubling each number to get the next, while the other column will take half the number and drop the remainder. We select the left-side column to be the doubling column, and the right-side column to be the halving column. Note that by halving the odd number such as 23 (the third number in the second column), we get 11 with a remainder of 1 and we simply drop the 1. The rest of this halving process should be clear.

43	92
86	46
172	23
344	11
688	5
1376	2
2752	1

Figure 1.6

Find the odd numbers in the halving column (here the right column) and then get the sum of the partner numbers in the doubling column (in this case the left column). These are highlighted in bold type. This sum gives you the originally required product of 43 and 92. In other words, with the Russian peasant's method, we get $43 \times 92 = 172 + 344 + 688 + 2752 = 3956$.

In the example above, we chose to have the left-side column, the doubling column, and the right-side column the halving column. We could also have done this Russian peasant's method by halving the numbers in the first column and doubling those in the second. See Figure 1.7.

43	92
21	184
10	368
5	736
2	1472
1	2944

Figure 1.7

To complete the multiplication, we find the odd numbers in the halving column (in bold type) and then get the sum of their partner numbers in the second column (now the doubling column). This gives us $43 \times 92 = 92 + 184 + 736 + 2944 = 3956$.

You are not expected to do your multiplication in this high-tech era by employing the Russian peasant's method. However, it should be fun to observe how this primitive system of arithmetic actually does work. Explorations of this kind are not only instructive but should also be entertaining.

Here you see what was done in the above multiplication algorithm:

$$
\begin{aligned}
{}^{*}43 \times 92 &= (21 \times 2 + 1)(92) &&= 21 \times 184 + &&92 = 3956 \\
{}^{*}21 \times 184 &= (10 \times 2 + 1)(184) &&= 10 \times 368 + &&184 = 3864 \\
10 \times 368 &= (5 \times 2 + 0)(368) &&= 5 \times 736 + &&0 = 3680 \\
{}^{*}5 \times 736 &= (2 \times 2 + 1)(736) &&= 2 \times 1472 + &&736 = 3680 \\
2 \times 1472 &= (1 \times 2 + 0)(1472) &&= 1 \times 2944 + &&0 = 2944 \\
{}^{*}1 \times 2944 &= (0 \times 2 + 1)(2944) &&= \quad 0 \quad + &&\underline{2944} = 2944 \\
& && &&3956
\end{aligned}
$$

Here in the next-to-last column, the numbers are added to get 3,956.

For those familiar with the binary system (i.e., base 2), one can also explain this Russian peasant's method with the following representation:

$$(43)(92) = [(1 \times 2^5) + (0 \times 2^4) + (1 \times 2^3) + (0 \times 2^2)$$

$$+ (1 \times 2^1) + (1 \times 2^0)](92)$$

$$= (2^0 \times 92) + (2^1 \times 92) + (2^3 \times 92) + (2^5 \times 92)$$

$$= 92 + 184 + 736 + 2944$$

$$= 3956$$

Whether or not you have a full understanding of the discussion of the Russian peasant's method of multiplication, you should at least, now, have a deeper appreciation for the multiplication algorithm you learned in school, even though most people today multiply with a calculator. There are many other multiplication algorithms, yet the one shown here is perhaps one of the strangest and it is through this strangeness that we can appreciate the powerful consistency of mathematics that allows us to conjure up such an algorithm.

Another Unusual Method to Multiply Two Numbers

Let's delve right into this unusual method for multiplying a pair of two-digit numbers with some examples, and then see why this technique works. Take, for example, the multiplication 95×97. The following steps can be done mentally (with some practice, naturally!):

Step 1: $95 + 97$ $=$ 192
Step 2: Delete the hundreds digit $=$ 92
Step 3: Tag two zeros onto the number $=$ 9,200
Step 4: $(100 - 95) \cdot (100 - 97)$ $=$ $5 \times 3 = 15$
Step 5: Add the last two numbers $=$ 9215, which is the
                                                                                      product being sought!

Here is another multiplication calculation using this technique to multiply a pair of two-digit numbers: $93 \times 96 = ?$

$93 + 96 = 189$

~~1~~89 (Delete the hundreds digit.)

Tag on two zeros $= 8,900$

Then add $(100 - 93) \times (100 - 96) = 7 \times 4 = 28$ to the previously obtained number to get 8,928, which is what we sought 93×96.

This technique also works when seeking the product of a pair of two-digit numbers that are further apart, such as $89 \times 73 =$?

$89 + 73 = 189$

~~1~~62 (Delete the hundreds digit.)

Tag on two zeros $= 6,200$

Then add $(100 - 89) \times (100 - 73) = 11 \times 27 = 297$, since this is a three-digit result, we add the hundreds digit, 2, to 62 and then follow the previous procedure to get 6,497, which again provides the product of 89×73.

For those who might be curious to know why this unusual technique works, we can use simple algebra to justify it. We begin with a pair of two-digit numbers:

$(100 - a)$ and $(100 - b)$ (where $0 < a$ and $b < 100$).

Step 1: $(100 - a) + (100 - b) = 200 - a - b$

Step 2: Delete the hundreds digit — which means subtracting 100 from the number: $(200 - a - b) - 100 = 100 - a - b$

Step 3: Tag on two zeros, which means multiply by 100:
$(100 - a - b) \times 100 = 10,000 - 100a - 100b$

Step 4: $a \times b$

Step 5: Add the last two results from Steps 3 and 4, and then do some factoring:

$$10,000 - 100a - 100b + ab$$

$$= 100(100 - a) - (100b - ab)$$

$$= 100(100 - a) - b(100 - a)$$

$$= (100 - a)(100 - b), \text{ which is what we set out to show.}$$

Now you just need to practice this technique to master it!

A Quick Multiplication Technique

There are people who are very adept at rapid multiplication. Here is one of the procedures that is used to do this quick multiplication of a pair of two-digit numbers. It is done as follows:

- *Begin by multiplying the units digits of the two numbers. If a two-digit number results, then write the units digit and carry the tens digit to the next step.*
- *Next, multiply the units digit of one number by the tens digit of the other number, and then multiply the tens digit of the first number by units digit of the second number and add the two products. Afterwards, add any number carried over from the previous step. Place the units digit of the result to the left of the previously obtained single digit and carry the tens digit to be added to the next step's calculation.*
- *Lastly, multiply the two tens-digits and add the carried-over number from the previous step. Place this result to the left of the two previously obtained numbers, and you will have your final result of the multiplication.*

At first sight, this appears to be a rather complicated procedure, but after we do an example, the procedure will become rather simple and clear and relatively fast to do. As our example, let us use this procedure to multiply 59×38.

- First, we will multiply the units digits of the two numbers to be multiplied: $9 \times 8 = 72$. This will have given us the units digit of our ultimate answer, namely, the **2**, and we will carry the 7 to the next step.
- We will now multiply the units and tens digits of the two numbers: $5 \times 8 = 40$ and $9 \times 3 = 27$. We now add these two numbers $40 + 27 = 67$ and then add the 7 carried from the previous step to get $67 + 7 = 74$. We place the **4** to the left of the previously obtained **2** and carry the 7 to the next step. So far, we have the tens and units digits of our sought-after product, namely, **42**.
- We now multiply the two tens digits to get $3 \times 5 = 15$ and add the carried 7 to get $15 + 7 = 22$, which we now place to the left of the two previously obtained final digits to get 2,242, which is the product of 59×38.

A quick review of this procedure is simply to multiply the pair of units digits, and the cross product of tens and units digits, and then the pair of tens digits — each time carrying the tens digit as appropriate. To be a successful user of this technique, you will need to practice this procedure with a number of examples so that your calculating speed will be impressive. An ambitious reader may want to extend this multiplication technique to multiplying two-digit numbers by three-digit numbers, and three-digit numbers by other three-digit numbers.

A Multiplication Technique for Special Numbers

Here is a technique for multiplying by two-digit numbers whose units digit is 1, such as 21, 31, and 41. Follow along step by step and eventually the technique will become easier.

To multiply by 21: *Double the number, then multiply by 10, and add the original number.*

For example: To multiply 37 × 21, double 37 yields 74, multiply by 10 to get 740, and then add the original number 37 to get 777.

Here's another example:

To multiply by 31: *Triple the number, then multiply by 10, and add the original number.*

For example: To multiply 43 × 31, triple 43 yields 129, multiply by 10 to get 1,290, and then add the original number 43 to get 1,333.

For some more practice, we offer the following:

To multiply by 41: *Quadruple the number, then multiply by 10, and add the original number.*

For example: To multiply 47 × 41, quadruple 47 yields 188, multiply by 10 to get 1,880, and then add the original number 47 to get 1,927.

By now you should be able to recognize the pattern and apply the rule to other such two-digit numbers.

This technique also works for multiplying by two-digit numbers whose units digit is a 9, such as 19, 29, and 39.

To multiply by 19: *Double the number, then multiply by 10, and subtract the original number.*

For example: To multiply 37 × 19, double 37 yields 74, multiply by 10 to get 740, and then subtract the original number 37 to get 703.

Although experiencing the solution to arithmetic problems can be useful when a computer or calculator is not at hand, considering the previously presented techniques provides a keen insight into mathematics, which could open the door for further arithmetic investigations.

Shopping with Mathematical Support

Most supermarkets today provide the unit cost of an item. This is very helpful as it allows the consumer to decide whether it makes sense to buy two 12 oz. jars of mayonnaise costing $1.35 per jar or one 30 oz. jar of the same brand of mayonnaise costing $3.49. We have been trained to think that the larger quantity is generally the better price value. However, there is a neat little trick to determining which is the better price per ounce when it isn't provided by the market. First, we need to establish the price per ounce for each of the two jars:

For the 12-ounce jars, the price per ounce is $\frac{\$1.35}{12}$

For the 30-ounce jar, the price per ounce is $\frac{\$3.49}{30}$

To compare the two fractions $\frac{1.35}{12}$, $\frac{3.49}{30}$ in order to see which is larger, there is a neat little algorithm to accomplish this task. We will cross multiply, writing the products under the fraction whose numerator was used (see Figure 1.8).

$$\frac{\$1.35}{12} \diagdown \diagup \frac{\$3.49}{30}$$

$$1.35 \times 30 \qquad 3.49 \times 12$$

$$40.50 \quad < \quad 41.88$$

Figure 1.8

The larger product, in this case 41.88, determines that the fraction $\frac{3.49}{30}$ is a larger fraction, and therefore, the more expensive cost of mayonnaise, in this case the larger jar was more expensive per ounce than the smaller jar. Although this is not typically expected, it does occur, and for that reason, a good consumer will make these comparisons.

Successive Percentages

We encounter challenges in our everyday lives and don't even realize that they can be properly understood with just a little mathematics. We often visit stores that are running a sale and then on a special day will *add* a percentage on top of the one having been advertised previously. The typical response is to add to percentages and conclude that the total savings for the day would be the sum of the two percentages. This is clearly a wrong calculation. Most folks defer thinking about percentage problems as they see them as nothing but a nemesis. Problems get particularly unpleasant when multiple percentages need to be processed in the same problem. However, we shall see how such successive percentages lend themselves very nicely to a delightfully simple arithmetic algorithm that leads us to lots of useful applications and provides new insight into successive percentage problems. This not-very-well-known procedure should be enriching.

Let's begin by considering the following problem:

> Wanting to buy a coat, Barbara is faced with a dilemma. Two competing stores next to each other carry the same brand coat with the same list price but with two different discount offers. Store A offers a 10% discount year-round on all its goods but on this particular day, offers an additional 20% on top of its already discounted price. Store B simply offers a discount of 30% on that day in order to stay competitive.
>
> Are the two end prices the same? If not, which gives Barbara the better price?

At first glance, you may assume there is no difference in price, since $10 + 20 = 30$, which would appear to be yielding the same discount in both cases. Yet with a little more thought you may realize that this is not correct, since in store A only 10% is calculated on the original list price, while the 20% discount is calculated on the lower price (namely, the 10% discounted price), while at store B, the entire 30% is calculated on the original price. Now, the question to be answered is as follows: What percentage difference is there between the discount in store A and store B?

To determine the difference in the prices, one procedure might be to assume the cost of the coat to be $100 and then calculate the 10% discount yielding a $90 price, and an additional 20% of the $90 price (or $18) will bring the price down to $72. In store B, the 30% discount on $100 would bring the price down to $70, giving a discount difference of $2 between the two stores, which in this case would be a 2% difference. This procedure, although correct and not too difficult, is a bit cumbersome and does not always allow a full insight into the situation as you will soon see.

We shall provide an interesting and quite unusual procedure for a deeper look at this situation, as well as for entertainment. We will consider a somewhat mechanical method for obtaining a single percentage discount (or increase) equivalent to two (or more) successive discounts (or increases). Follow this four-step procedure:

(1) *Change each of the percents involved into decimal form:*
    .20 and .10
(2) *Subtract each of these decimals from 1.00:*
    .80 and .90 (for a percent increase, add to 1.00)
(3) *Multiply these decimals:*
    .80 × .90 = .72
(4) *Subtract this number from 1.00:*
    1.00 − 0.72 = 0.28, which, written as a percentage is 28%, represents the combined discount.
    If the result of Step 3 is greater than 1.00, subtract 1.00 from it to obtain the percent of increase.

Therefore, we can conclude that the combined percentage of 28% differs from the single discount of 30% by 2%.

Following the same procedure, you can also combine more than two successive discounts. In addition, successive increases, combined or not combined with a discount, can also be accommodated in this procedure by adding the decimal equivalent of the increase to 1.00, while the discount was subtracted from 1.00 and then continuing the procedure in the same way. If the end result comes out greater than 1.00, then this will have resulted in an overall increase rather than the discount as found in the above problem.

A conundrum often facing consumers is that of determining if a discount and increase of the same percentage leave the original price unchanged. For example, suppose a store just increased all of its prices by 10% and then notes that its business has declined substantially, whereupon they then resort to discounting all of these recently increased prices by the same percentage of 10%. Have they then restored the prices to their original level? Using this technique, we find ourselves multiplying 1.1 times 0.90 to get 0.99, which would indicate that the original price had dropped by 1%. For many people, this is a counterintuitive result.

This procedure not only streamlines a typically cumbersome situation but also provides some insight into the overall picture. For example, consider the following question: Is it advantageous to the buyer in the above problem to receive a 20% discount and then a 10% discount, or the reverse, 10% discount and then a 20% discount? The answer to this question is not immediately intuitively obvious. Yet, since the procedure just presented shows that the calculation is merely multiplication, a commutative operation, we can immediately conclude that there is no difference between the two.

So here you have a delightful algorithm for combining successive discounts or increases or combinations of these to calculate the combined result. Not only is it useful, but also it gives you some newfound power in dealing with percentages when a calculator might not be available.

Another shopping situation where mathematics can be helpful is when there are discounts of different types. Suppose you have two sales-promotion coupons for the same store: One that says "20% off" (independent of the purchase) and one that only applies above a specified amount, for example, "$15 off for purchases exceeding $49.99." Assuming that the two coupons cannot be combined, then which one would be more advantageous, if the item we want to buy costs, say, $80? The 20% off coupon would yield a price of $64, while the $15 reduction would yield a price of $65. It might be nice to know at what price the 20% coupon will be more advantageous.

To approach this problem, we may consider two extreme cases. Let's consider an item costing $50, since the minimal purchase that qualifies for the second coupon is $50, which when reduced by $15 would be $35. On the other hand, the fixed-percentage coupon would yield a reduction of 20%, or one-fifth, of $50, which is $10. That means we would have to pay $40, if we used the fixed-percentage coupon. Thus, the $15 coupon would be the better choice for a $50 purchase.

The other extreme case could be infinity, but since we can only spend a finite amount of money, let's assume our budget is very high, say it is $150. One-fifth of $150 is $30. Therefore, the 20% coupon yields a reduced price of $120, while the other coupon (a reduction of $15) would result in a price of $135. Obviously, the better choice of coupon depends on the total sum of the purchase. For the fixed-percentage coupon, the amount of money saved increases with the price, while we cannot save more than $15 with the other coupon. The extreme cases we have considered show that there must be some break-even point X between $50 and $150, at which price, both coupons yield the same discount. For purchases at $50, where the 20% reduction will be $10 so that the $15 is preferred, yet, for purchases above $50 and below X, the $15 coupon is still the better choice. To find the break-even price X for a 20% coupon and a $15 coupon, we just have to compute $X - .20X = X - \$15$ and then $X = \$75$. Therefore, if we want to buy an item for more than $75, we should use the 20% coupon.

Occasionally, you may also encounter different-type coupons that are combinable, although this is a rare phenomenon since most stores are usually not that generous to their customers. Let's take a look at such a situation, since it provides an example for the mathematical notion of non-commutativity. Suppose we were allowed to use both coupons for the same purchase, that is, coupon 1 with a 20% discount as well as coupon 2 with a $15 reduction. Now the question arises whether the order matters or not. If it were to matter, which of the two coupons should be used first? Denoting the price without any discount by P (which we assume to be at least $50 for the sake of simplicity), we obtain the following:

- A reduced price by using coupon 1 first as $p_{1,2} = P \cdot 0.8 - \$15$ and;
- A reduced price by using coupon 2 first as $p_{2,1} = (P - \$15) \cdot 0.8$.

Since $p_{2,1} = (P - \$15) \cdot 0.8 = P \cdot 0.8 - \$12$, which is more than $p_{1,2}$. Therefore, we should apply coupon 1 first, unless $P \cdot 0.8$ is less than $50. In this case, we would have to use coupon 2 first.

Two operations that will in general lead to different results, if their order is reversed, are called "non-commutative" in mathematics. As our analysis showed, these different types of discounts are an example of non-commutative operations, meaning that the order does matter! You should think about that, if you are offered combinable discounts of different types. However, requesting your preferred order might be a bit of a challenge.

Raising Interest

We are often confronted with advertisements by savings institutions offering attractive interest rates and frequent compounding of interest on deposits. Since most banks have a variety of programs, it is valuable for potential depositors to understand how interest is calculated under each of the available options. In our discussion of interest rates and practices, we will use the formula for compound interest to calculate the return on investments at any rate of interest, for any period of time, and for any commonly used frequency of compounding, including instantaneous (continuous) compounding. They will also determine which of the two or more alternatives gives the best return over the same time period.

Let's consider the following interesting problem:

In the year 1626, Peter Minuit bought Manhattan Island for the Dutch West India Company from the Lenape Native Americans for trinkets costing 60 Dutch guilders or about $24. Suppose the Lenapes had been able to invest this $24 at that time at an annual interest rate of 6%, and suppose further that this same interest rate had continued in effect all these years. How much money could the present-day descendants of these original Lenapes collect if (1) only simple interest was calculated and (2) interest was compounded (a) annually, (b) quarterly, and (c) continuously?

The answers to these questions could be surprising.

You might recall that simple annual interest is calculated by taking the product of the principal P, the annual interest rate r, and the time in years t. Accordingly, you have the formula $I = Prt$, and in the above problem, $I = (24)(.06)(397) = \$571.68$ as simple interest. Add this to the principal of $24.00 to obtain the amount A of $595.68 available in 2023. You have just used the formula for "amount," $A = P + Prt$.

With this relatively small sum in mind (for a return after 397 years!), let's investigate the extent to which this return would have been improved if interest had been compounded annually instead of being calculated on only a simple basis. With a principal P, an annual rate of interest r, and a time $t = 1$, the amount A at the end of the first year is given by the formula $A_1 = P + Pr = P(1 + r)$. (The subscript indicates the year at the end of which interest is calculated.) Now $A_1 = P(1 + r)$ becomes the principal at the beginning of the second year, upon which interest will be credited during the second year.

Therefore, $A_2 = P(1+r) + P(1+r)r = P(1+r)(1+r) = P(1+r)^2$. Since the last expression represents the principal at the beginning of the third year, you have

$$A_3 = P(1+r)^2 + P(1+r)^2r = P(1+r)^2(1+r) = P(1+r)^3.$$

By now, you will see the emerging pattern and should be able to suggest the generalization for the amount after t years, $A_t = P(1+r)^t$.

Now try this formula on the $24 investment made in 1626. Assuming annual compounding at 6% per annum, you have $A_{397} = 24(1 + .06)^{397} = 267,079,006,366$. This means that the original $24 is now worth more than $200 billion! One should be surprised by the huge difference between this figure and the $595.68 obtained by computing simple interest.

Most banks now compound not annually but quarterly, monthly, daily, or continuously, so we shall next generalize the formula $A = P(1 + r)^t$ to take into account compounding at more frequent intervals. Bear in mind that if interest is compounded semi-annually, the *periodic rate* would be only *one-half* the annual rate, but the number of periods would be *twice* the number of years so that $A = P\left(1 + \frac{r}{2}\right)^{2t}$. Likewise, if the interest is compounded quarterly, we have $A = P\left(1 + \frac{r}{4}\right)^{4t}$. In general, if the interest is compounded n times a year, the formula would be $A = P\left(1 + \frac{r}{n}\right)^{nt}$, which can be used for any finite value of n. Letting $n = 4$ in our original problem yields $A = 24\left(1 + \frac{.06}{4}\right)^{4(397)} = 24(1.015)^{1588} = 444,924,512,737$. Our $24 has now risen to about $445 billion. Note that changing the compounding from annually to quarterly increased the yield by about $178 billion. You may now wonder whether the yield can be increased indefinitely by simply increasing the frequency of compounding. A complete treatment of this question requires a thorough development of the concept of limits, but an informal, intuitive approach will suffice here. We shall first explore the simpler problem of an investment of $1 at a nominal annual interest rate of 100% for a period of one year. This will lead us to $A = 1\left(1 + \frac{100}{n}\right)^n$. You should now prepare a table of values for A for various common values of n, such as $n = 1$ (annual compounding), $n = 2$ (semiannual), $n = 4$ (quarterly), and $n = 12$ (monthly). Note that the amount A does *not* rise astronomically as n increases but rather rises slowly from $2.00 ($n = 1$) to about $2.60 ($n = 12$). The amount A would approach, but not quite reach, the value $2.72. To explain this limiting value takes us a bit beyond the realm of this book. Yet for the more advanced reader, it is $\lim_{n \to \infty} \left(1 + \frac{1}{n}\right)^n = e = 2.71828...$ This number e is called "Euler number".

Since investments generally do not earn 100% interest, we must convert to a general interest rate, say r. By letting $\frac{r}{n} = \frac{1}{k}$, we have $n = kr$, and $A = P\left(1 + \frac{r}{n}\right)^{nt}$, which then becomes $A = P\left(1 + \frac{1}{k}\right)^{krt} = P\left[\left(1 + \frac{1}{k}\right)^k\right]^{rt}$. Clearly, as n approaches infinity so does k, since r is finite, so the expression in brackets approaches the value e as a limit. You then have the formula $A = Pe^{rt}$ for instantaneous compounding, where r is the nominal annual rate of interest and t is the time in years. You might be interested in knowing that this formula is a special representation of the general "Law of Growth," which is usually written in the $N = N_0e^{rt}$ form, where N represents the final amount of a material whose initial amount was N_0. This law has applications in many other areas such as population growth (of people, bacteria in a culture, etc.) and the radioactive decay of elements, in which case it becomes the "law of decay," as $N = N_0e^{-rt}$.

Completing the investment problem, using 2.72 as an approximation to e, you have $A = 24(2.72)^{.06(397)} = 539{,}066{,}738{,}490$. So, you can see that the "ultimate" return on a $24 investment (at a nominal annual interest rate of 6% for 397 years) is over $500 billion.

Banks currently offer much lower interest rates and compounding is commonly done quarterly, monthly, daily, or continuously. You can work problems with varying principals, periodic rates, frequencies of compounding, and time periods and compare yields. You may be surprised at the outcomes.

The Rule of 72

We often want to know how a certain interest rate in a bank will affect our total holdings. Naturally, there are traditional ways of calculating interest, which we have just experienced in the previous unit. However, there is an unusual quirk of our number system that allows us to calculate how long it will take to double our money in a bank with a yearly compounding procedure at any given annual percentage rate. The procedure is as fast as you can divide 72 by another number. This is clearly good to know, but it is the unusualness of this rule that allows us to exhibit it here. So, enjoy it. It is called the "Rule of 72" as it is based on this number as you will soon see.

The "Rule of 72" states that, roughly speaking:

Money will double in $\frac{72}{r}$ years when it is invested at an annual com-pounded interest rate of r%.

So, for example, if we invest money at an 8% compounded annual inter-est rate, it will double its value in $\frac{72}{8} = 9$ years. Similarly, if we leave our money in the bank at a compounded rate of 2%, it would take 36 years for this sum to double its value. The interested reader might want to bet-ter understand why this is so, and how accurate it really is. The following discussion will explain that.

To investigate why or if this really works, we consider the compound interest formula: $A = P\left(1 + \frac{r}{100}\right)^n$, where A is the resulting amount of money and P is the principal invested for n interest periods at $r\%$ annually. We need to investigate what happens when $A = 2P$.

The above equation then becomes $2 = \left(1 + \frac{r}{100}\right)^n$ (1)

It then follows that $n = \frac{\log 2}{\log\left(1 + \frac{r}{100}\right)}$ (2)

A table of values (Figure 1.9) from the above equation established with the help of a calculator is shown.

If we take the arithmetic mean (the usual average) of the nr values, we get 72.04092314, which is quite close to 72, and so our "Rule of 72" seems to be a very close estimate for doubling money at an annual interest rate of $r\%$ for n interest periods.

If r is restricted to more realistic values between, say, .5% and 5%, then one could use also the "Rule of 70."

r	n	nr
1	69.66071689	69.66071689
3	23.44977225	70.34931675
5	14.20669908	71.03349541
7	10.24476835	71.71337846
9	8.043231727	72.38908554
11	6.641884618	73.0607308
13	5.671417169	73.72842319
15	4.959484455	74.39226682

Figure 1.9

An ambitious reader, or one with a very strong mathematics background, might try to determine a "rule" for tripling and quadrupling money, similar to the way we dealt with the doubling of money. The above equation (2) for k-tupling would be $n = \frac{\log k}{\log\left(1 + \frac{r}{100}\right)}$, which for $r = 8$, gives the value for $n = 29.91884022(\log k)$. Thus $nr = 239.3507218(\log k)$, which for $k = 3$ (the tripling effect) gives us $nr = 114.1993167$. We could then say that for tripling money, we would have a "Rule of 114." For $k = 4$, we even do not need any further calculations. Since two times doubling is equivalent to quadrupling, we know that here we would have a "Rule of 144" (or "Rule of 140" if we used the "Rule of 70" for doubling). Although we use computer calculation, for the most part, to determine compound interest, here we see a simple trick that allows us to skip sizable portion of calculation and still come up with a useful answer. Once again, we see how mathematics helps us with everyday finances.

Chapter 2

Numerical Novelties

Mathematics harbors a plethora of numerical curiosities. There are times when one simply stumbles on some of these and there are other times when one seeks a creative challenge to find such curiosities. In this chapter, we present some of these unusual numerical relationships with the hope that they will not only bring a new life into one's view of mathematics, but also perhaps motivate readers to seek other analogous relationships. We begin with a few simple ones and then progress further to some more challenging, yet amazing, numerical novelties.

The Nine-Digit Number Yields an Unexpected Digit Sum

This sum needs no further explanation. Just admire it!

$$
\begin{array}{ll}
987,654,321 & \rightarrow \text{digit sum } 45 \\
-123,456,789 & \rightarrow \text{digit sum } 45 \\
\hline
864,197,532 & \rightarrow \text{digit sum } 45
\end{array}
$$

Curiously, 16 Fours Can Equal 1,000

Here, the number 4 is used 16 times to create the number 1,000.

$$444 + 444 + 44 + 44 + 4 + 4 + 4 + 4 + 4 + 4 = 1,000$$

Unexpectedly, 8 Eights Can Equal 1,000

Here, the digit 8 is used 8 times to create the number 1,000.

$$888 + 88 + 8 + 8 + 8 = 1,000$$

Using Nine Digits to Create the Number One-Half

Note how all 9 digits are used exactly once to create the number $\frac{1}{2}$.

$$\frac{9,327}{18,654} = \frac{1}{2}$$

Using Nine Digits to Create 99,999

Here is an addition calculation that uses all 9 digits exactly once with a sum of 99,999:

$$\begin{array}{r} 98,765 \\ +1,234 \\ \hline 99,999 \end{array}$$

An Unexpected Pattern

Patterns in mathematics tend to crop up when we would least expect them. We offer the following example of how a rather innocuous series of numbers, when squared, results in a totally unexpected pattern.

$$4^2 = 16$$
$$34^2 = 1156$$
$$334^2 = 111,556$$
$$3334^2 = 11,115,556$$
$$33334^2 = 1,111,155,556$$
$$333334^2 = 111,111,555,556$$

Although the ambitious reader will probably be able to continue this pattern, it would be pleasantly challenging to seek other patterns of a similar kind. Such unusual occurrences in mathematics are always entertaining as well as motivating.

An Unexpected Prime Relationship

Here, we encounter an interesting phenomenon in arithmetic, where we begin by choosing any sequence of natural numbers, beginning with 1 and ending before it reaches the next prime number. Then, multiply those numbers and add 1 to this product, and you will find that this result is divisible by the prime number at which you stopped the sequence.

To show how this works, let's consider the first few such sequences that you might select, such as the sequence that ends before the prime number 3, and perform the requested operation:

$$1 \times 2 + 1 = 3, \text{ which we find is divisible by 3}$$

Now, consider the sequence that ends before the next prime number, namely, 5:

$$1 \times 2 \times 3 \times 4 + 1 = 25, \text{ which is divisible by 5}$$

Considering the sequence that ends before the next prime number, namely, 7:

$$1 \times 2 \times 3 \times 4 \times 5 \times 6 + 1 = 271, \text{ which is divisible by 7, since } \tfrac{721}{7} = 103$$

Taking this one step further to the next prime number, namely, 11:

$$1 \times 2 \times 3 \times 4 \times 5 \times 6 \times 7 \times 8 \times 9 \times 10 + 1 = 3{,}628{,}801,$$

$$\text{which is divisible by 11, since } \frac{3{,}628{,}801}{11} = 329{,}891$$

This then continues on and once again demonstrates amazing numerical relationships.

Favorable Numerical Arrangements of Digits 1–9

The challenge here is to arrange the numbers 1, 2, 3, 4, 5, 6, 7, 8, 9 in sequence using only addition and subtraction to reach the number 100. Here is one possible solution:

$$123 - 45 - 67 + 89 = 100$$

We are offering some other solutions that you might get from your audience just so you will be properly prepared:

$$123 + 4 - 5 + 67 - 89 = 100$$
$$123 + 45 - 67 + 8 - 9 = 100$$
$$123 - 4 - 5 - 6 - 7 + 8 - 9 = 100$$
$$12 - 3 - 4 + 5 - 6 + 7 + 89 = 100$$
$$12 + 3 + 4 + 5 - 6 - 7 + 89 = 100$$
$$1 + 23 - 4 + 5 + 6 + 78 - 9 = 100$$
$$1 + 2 + 34 - 5 + 67 - 8 + 9 = 100$$
$$12 + 3 - 4 + 5 + 67 + 8 + 9 = 100$$
$$1 + 23 - 4 + 56 + 7 + 8 + 9 = 100$$
$$1 + 2 + 3 - 4 + 5 + 6 + 78 + 9 = 100$$
$$-1 + 2 - 3 + 4 + 5 + 6 + 78 + 9 = 100$$

To take this one step further, we can also try to do this in reverse, such as

$$9 + 8 + 76 + 5 - 4 + 3 + 2 + 1 = 100$$

Here are several more examples of how to create sums of 100 using the numbers 1–9 in reverse order:

$$98 - 76 + 54 + 3 + 21 = 100$$
$$9 - 8 + 76 + 54 - 32 + 1 = 100$$
$$98 - 7 - 6 - 5 - 4 + 3 + 21 = 100$$
$$9 - 8 + 7 + 65 - 4 + 32 - 1 = 100$$

$$9 - 8 + 76 - 5 + 4 + 3 + 21 = 100$$

$$98 - 7 + 6 + 5 + 4 - 3 - 2 - 1 = 100$$

$$98 + 7 - 6 + 5 - 4 + 3 - 2 - 1 = 100$$

$$98 + 7 + 6 - 5 - 4 - 3 + 2 - 1 = 100$$

$$98 + 7 - 6 + 5 - 4 - 3 + 2 + 1 = 100$$

$$98 - 7 + 6 + 5 - 4 + 3 - 2 + 1 = 100$$

$$98 - 7 + 6 - 5 + 4 + 3 + 2 - 1 = 100$$

$$98 + 7 - 6 - 5 + 4 + 3 - 2 + 1 = 100$$

$$98 - 7 - 6 + 5 + 4 + 3 + 2 + 1 = 100$$

$$9 + 8 + 76 + 5 + 4 - 3 + 2 - 1 = 100$$

$$-9 + 8 + 76 + 5 - 4 + 3 + 21 = 100$$

$$-9 + 8 + 7 + 65 - 4 + 32 + 1 = 100$$

$$-9 - 8 + 76 - 5 + 43 + 2 + 1 = 100$$

You might also extend your newly developed talent by creating the number 100 using all 10 digits with only the operations of addition and multiplication. One such possibility using all digits from 0 to 9 is $(9 \times 8) + 7 + 6 + 5 + 4 + 3 + 2 + 1 + 0$.

Using All Nine Digits 1–9 as Mixed-Number Fractions

Here, we try to find a way of representing the number 100 using all nine digits in the form of a mixed-number fraction. We offer 11 ways to do that and they are by no means simple, but perhaps after seeing a few of them, you may discover other ways to accomplish this feat. Here are the 11 possibilities:

$$3\frac{69258}{714}, \quad 81\frac{5643}{297}, \quad 81\frac{7524}{396}, \quad 82\frac{3546}{197}, \quad 91\frac{5742}{638}, \quad 91\frac{5823}{647},$$

$$91\frac{7524}{836}, \quad 94\frac{1578}{263}, \quad 96\frac{1428}{357}, \quad 96\frac{1752}{438}, \quad 96\frac{2148}{537}$$

More Fun with All 10 Digits 0–9

Here, we seek a way to arrange the digits 0–9 in fraction form to reach the number 1. One possible solution is $\frac{35}{70} + \frac{148}{296} = \frac{1}{2} + \frac{1}{2} = 1$.

Another challenge could be to create the number 10 using all 10 digits. One possible way to create the number 10 by using all digits from 0 to 9 exactly once is $1\frac{35}{70} + 8\frac{46}{92} = 10$.

An Amazing Result by Multiplying 12,345,679 by a Multiple of 3

Here, we can have some fun by multiplying the number 12,345,679 (note that 8 is missing) by various multiples of 3 to get some surprising results. Keeping in mind that the numbers 45, 48, and 51 are all multiples of 3 and so we offer the following examples:

$$12345679 \times 45 = 555,555,555$$

$$12345679 \times 48 = 592,592,592$$

$$12345679 \times 51 = 629,629,629$$

Here are some more examples using other multiples of 3, such as 63, 54, and 72:

$$12345679 \times 63 = 777,777,777$$

$$12345679 \times 54 = 666,666,666$$

$$12345679 \times 72 = 888,888,888$$

You might wonder, which multiples of 3 resulted in numbers which had a repetition of 9 digits. There are many other surprising results through these multiple-of-3 multiplications. Each will probably bring surprise and pleasure.

A Surprising Division

Today, it is quite likely that you have a calculator on hand which will allow you to experience a rather pretty result, namely, where our 10 digits keep repeating as shown in the following (only the first three repetitions; therefore, we use the symbol $\approx$ instead of $=$):

$$\frac{137,174,210}{1,111,111,111} \approx 0.\mathbf{1234567890}\,123456789 0\mathbf{1234567890}$$

This beautiful result requires no further explanation.

Returning to a Starting Number

Select any three-digit number and write it twice to form a six-digit number. For example, if you choose the number 357, then write the six-digit number 357,357. We now — perhaps using a calculator — divide this number by 7, then divide the resulting quotient by 11, and lastly, divide that quotient by 13, as we have done as follows:

$$\frac{357,357}{7} = 51,051$$

$$\frac{51,051}{11} = 4,641$$

$$\frac{4,641}{13} = 357$$

You will recognize 357 as the first number we started with! You may wish to try this with other three-digit numbers to convince yourself that this technique always works.

Explanation: The reason that this works is that to form the original six-digit number, you actually multiplied the original three-digit number by 1,001. That is, $357 \times 1,001 = 357,357$. However, $1,001 = 7 \times 11 \times 13$. Therefore, by dividing successively by 7, 11, and 13, we have undone the original multiplication by 1,001, leaving the original number. To those unaware of this relationship, the result will surely be surprising.

While we are on the subject of the number 1,001, we can see how this number can also help us multiply other number combinations:

$$221 \times 77 = (17 \times 13) \times (11 \times 7) = 1,001 \times 17$$

$$= 1,000 \times 17 + 1 \times 17 = 17,017$$

$$264 \times 91 = (24 \times 11) \times (13 \times 7) = 1,001 \times 24$$

$$= 1,000 \times 24 + 1 \times 24 = 24,024$$

$$407 \times 273 = (37 \times 11) \times (3 \times 7 \times 13) = 1,001 \times 111$$

$$= 1,000 \times 111 + 1 \times 111 = 111,111$$

Arithmetic can indeed expose some hidden numerical treasures.

Products of 91 and the Numbers 1–9

To appreciate the next suggested multiplications with the number 91, we need to do all the multiplications from 1 through 9 and then admire the results *vertically*. Note the list of units digits, tens digits, and hundreds digits:

$$91 \times 1 = 091$$

$$91 \times 2 = 182$$

$$91 \times 3 = 273$$

$$91 \times 4 = 364$$

$$91 \times 5 = 455$$

$$91 \times 6 = 546$$

$$91 \times 7 = 637$$

$$91 \times 8 = 728$$

$$91 \times 9 = 819$$

A Strange Coincidence

Can you imagine that the first six prime numbers could be divisors of six consecutive numbers? Well, there is such a case, where the consecutive numbers, 788, 789, 790, 791, 792, and 793, are divisible by the first six prime numbers: 2, 3, 5, 7, 11, and 13, respectively. That is,

$$\frac{788}{2} = 394, \quad \frac{789}{3} = 263, \quad \frac{790}{5} = 158, \quad \frac{791}{7} = 113,$$

$$\frac{792}{11} = 72, \quad \frac{793}{13} = 61$$

How to Establish a Prime Number

There are times when you will appreciate a clever technique for doing something that could have been taught in high school but clearly was not. One such case is how to establish whether a number is a prime number or not. Before we set up a technique for establishing prime numbers, we need to review the meaning of a factorial, such as $n! = 1 \times 2 \times 3 \times 4 \times 5 \times 6 \times \cdots \times n$. The rule for establishing a prime is that if $n! + 1$ is divisible by $n + 1$, then $n + 1$ is a prime number. Suppose we would like to test this to see if the number 11 is a prime number. Therefore, we say that $11 = n + 1$, whereupon $n = 10$. We now seek $10!$, which is equal to 3,628,800. Thus, $3,628,800 + 1 = 3,628,801 = 11 \times 329,891$. Therefore, we can conclude that 11 is a prime number. Although this may appear to be somewhat complicated, it is much simpler than taking a large number and trying to divide it by so many various smaller numbers until you can conclude that it has no factors other than itself and 1.

Explanation: Why does this work? This is part of the so-called *Wilson's theorem*. One direction of this famous number theorem (the one we used above) is easy to prove without a deep knowledge of number theory: The product $1 \times 2 \times 3 \times 4 \times 5 \times 6 \times \cdots \times n + 1$ cannot be divisible by $2, 3, \ldots, n$ (because the first summand $1 \times 2 \times 3 \times 4 \times 5 \times 6 \times \cdots \times n$ is divisible by these numbers. In this case, 1 would have to be divisible by these numbers, which is obviously not the case). Thus, $n + 1$ is the *smallest* number greater than 1 which can possibly be a divisor of $n + 1$, and the smallest divisor greater than 1 of a number must obviously be a prime number.

Twin Primes

Prime numbers can also be seen by their place on the list of primes. When two prime numbers have a difference of 2, they are considered *twin primes*. It is suspected that there are infinitely many twin primes, but this has never been proved or disproved. The first few twin primes are (3, 5), (5, 7), (11, 13), (17, 19), (29, 31), (41, 43), (59, 61), (71, 73), (101, 103), (107, 109), (137, 139)…. We note that 5 is the only number that will appear twice in the list of twin primes. You may be curious to know what the largest twin prime pair that has been discovered to date. As of June 2023, the largest twin prime pair is ($2996863034895 \times 2^{1290000} - 1$, $2996863034895 \times 2^{1290000} + 1$), with 388,342 decimal digits, which was discovered in September 2016.

You may ask if there is a general format for expressing twin prime pairs. The answer is that — with the exception of the first twin prime pair, namely, (3, 5) — they can be expressed in the form of $(6n - 1)$, $(6n + 1)$, where n is a natural number. A clever observer will also note that every number between a pair of twin primes will be a multiple of 6, which can be easily substantiated by examining the first several twin primes.

The Sum of a Two-Digit Number and its Reversal

The sum of a two-digit number and its reversal is always divisible by 11. As an example, consider the number $89 + 98 = 187$, which is divisible by 11.

Explanation: Now, consider the number sum $ab + ba$, which, when written out properly, is $(10a + b) + (10b + a) = 11a + 11b = 11(a + b)$, making 11 a factor of the number we started with.

The Numbers 9 and 11

The numbers 9 and 11 have very peculiar properties in the base 10 system since they are on either side of the base number 10. For example, in the "cubic world," $11^3 = 1331$, a palindromic number, and the number 9 also produces a very unusual pattern since $9^3 = 729 = 1^3 + 6^3 + 8^3 = 3^6$. An ambitious reader may wish to search for other analogous patterns.

Further Peculiarities with the Numbers 9 and 11

The two numbers, 9 and 11, can also be tied together in other ways. Such as $\frac{1}{9} = 0.111111111111111111111\ldots$ and $\frac{1}{11} = 0.0909090909090909\ldots$. We know that the product of 9 and 11 is 99 and that also provides a rather unusual unit fraction: $\frac{1}{99} = 0.0101010101010101\ldots$. While on the topic of 99, here is another curiosity that can be entertaining: $99^2 = 9801$, and if we split and add, $98 + 01 = 99$.

Taking the number 9 a step further, we can consider the number 999, which is the product of 27×37. Now, taking the reciprocals of these two numbers, another nice pattern and relationship evolves $\frac{1}{27} = 0.037037037037037037\ldots$ and when we compare that to $\frac{1}{37} = 0.027027027027027027\ldots$, we note a fascinating relationship between these two numbers that came from the factors of 999.

Not wanting to ignore 9's partner, the number 11 also provides some curious patterns, such as

$$11 = 6^2 - 5^2$$
$$1111 = 56^2 - 65^2$$
$$111,111 = 556^2 - 445^2$$
$$11,111,111 = 5556^2 - 4445^2$$

Multiplying Our Numbers by Multiples of 9

The number 9 and its multiples can produce a rather unexpected pattern of results when multiplied by a number formed with our 9 numerals in descending order. Once again, the calculator may be required to make this calculation run smoothly. It requires quite a bit of multiplication before the beauty of the pattern emerges.

Multiplying 987654321 by multiples of 9,

$$987654321 \times 9 = 8888888889$$

$$987654321 \times 18 = 17777777778$$

$$987654321 \times 27 = 26666666667$$

$$987654321 \times 36 = 35555555556$$

$$987654321 \times 45 = 44444444445$$

$$987654321 \times 54 = 53333333334$$

$$987654321 \times 63 = 62222222223$$

$$987654321 \times 72 = 71111111112$$

$$987654321 \times 81 = 80000000001$$

$$987654321 \times 90 = 88888888890$$

$$987654321 \times 99 = 97777777779$$

$$987654321 \times 108 = 106666666668$$

$$987654321 \times 117 = 115555555557$$

$$987654321 \times 126 = 124444444446$$

$$987654321 \times 135 = 133333333335$$

$$987654321 \times 144 = 142222222224$$

$$987654321 \times 153 = 151111111113$$

$$987654321 \times 162 = 160000000002$$

$$987654321 \times 171 = 168888888891$$

$$987654321 \times 180 = 177777777780$$

$$987654321 \times 189 = 186666666669$$

$$987654321 \times 198 = 195555555558$$

$$987654321 \times 207 = 204444444447$$

$$987654321 \times 216 = 213333333336$$

$$987654321 \times 225 = 222222222225$$

$$987654321 \times 234 = 231111111114$$

$$987654321 \times 243 = 240000000003$$

$$987654321 \times 252 = 248888888892$$

$$987654321 \times 261 = 257777777781$$

$$987654321 \times 270 = 266666666670$$

$$987654321 \times 279 = 275555555559$$

$$987654321 \times 288 = 284444444448$$

$$987654321 \times 297 = 293333333337$$

$$987654321 \times 306 = 302222222226$$

An ambitious reader may wish to continue to watch this pattern grow impressively. Patterns such as this, which are completely unexpected, serve well to exhibit the hidden beauty of mathematics.

Yet, we can take this a step further by reversing the order of the consecutive digits — without the number 8 — and multiplying by multiples of 9 and discover how another surprising pattern of numbers emerges:

$$12345679 \times 9 = 111,111,111$$

$$12345679 \times 18 = 222,222,222$$

$$12345679 \times 27 = 333,333,333$$

$$12345679 \times 36 = 444,444,444$$

$$12345679 \times 45 = 555,555,555$$

$$12345679 \times 54 = 666,666,666$$

$$12345679 \times 63 = 777,777,777$$

$$12345679 \times 72 = 888,888,888$$

$$12345679 \times 81 = 999,999,999$$

More Strange Multiples of 9

Multiplication with large numbers has become trivial using the calculator. However, some numbers have unusual properties and remain so after multiplication, especially when multiplied by the number 9, as we have seen in the previous example. It might be challenging to find numbers that are composed of eight different numerals that when multiplied by 9, result in numbers consisting of nine different numerals.

This can be an interesting challenge. We offer a few examples here:

$$58132764 \times 9 = 523194876$$

$$76125483 \times 9 = 685129347$$

$$72645831 \times 9 = 653812479$$

$$81274365 \times 9 = 731469285$$

What is particularly notable here is that the numeral 9 is missing from each number being multiplied by 9 and yet appears in each of the end products. Had we multiplied these original numbers by 18, we would be able to get numbers with 10 different digits; however, this time, the 0 would be included, which was excluded from the above products. Here are the products, each consisting of 10 different digits:

$$58132764 \times 18 = 1046389752$$

$$76125483 \times 18 = 1370258694$$

$$72645831 \times 18 = 1307624958$$

$$81274365 \times 18 = 1462938570$$

As neat as this appears, it still would be a challenge to find other such eight-digit numbers that when multiplied by 9, result in products that have 9 different digits, and when multiplied by 18, have products with 10 different digits.

More Fun with 9s

There are times when you can be amused with some unusual patterns in mathematics. The 9s lend themselves nicely to such a situation as you can see here.

$$9 \times 9 = 81$$
$$99 \times 99 = 9{,}801$$
$$999 \times 999 = 998{,}001$$
$$9{,}999 \times 9{,}999 = 99{,}980{,}001$$
$$99{,}999 \times 99{,}999 = 9{,}999{,}800{,}001$$
$$999{,}999 \times 999{,}999 = 999{,}998{,}000{,}001$$
$$9{,}999{,}999 \times 9{,}999{,}999 = 99{,}999{,}980{,}000{,}001$$

We also get an unexpected pattern of numbers when reverse consecutive-digit numbers are multiplied by 9 and are sequentially added to reverse consecutive numbers as shown here:

$$0 \times 9 + 8 = 8$$
$$9 \times 9 + 7 = 88$$
$$98 \times 9 + 6 = 888$$
$$987 \times 9 + 5 = 8{,}888$$
$$9{,}876 \times 9 + 4 = 88{,}888$$
$$98{,}765 \times 9 + 3 = 888{,}888$$
$$987{,}654 \times 9 + 2 = 8{,}888{,}888$$
$$9{,}876{,}543 \times 9 + 1 = 88{,}888{,}888$$
$$98{,}765{,}432 \times 9 + 0 = 888{,}888{,}888$$

This time, we take multiples of 9 and multiply them by the number 37,037, as shown in the following, to once again get some surprising results:

$$37{,}037 \times 3 = 111{,}111$$
$$37{,}037 \times 6 = 222{,}222$$
$$37{,}037 \times 9 = 333{,}333$$
$$37{,}037 \times 12 = 444{,}444$$
$$37{,}037 \times 15 = 555{,}555$$
$$37{,}037 \times 18 = 666{,}666$$

A Program for Determining the Number of Digits Used

A computer program prints in sequence the natural numbers 1, 2, 3, 4, 5, 6, 7, 8, 9, 10, 11, 12, 13, 14, 15, …. When we reach the number 15, there were 21 digits that have been printed. What is the largest natural number that can be entirely printed by this computer program generating consecutive numbers, if it can print at most 1,000,000,000 digits?

Explanation: There are nine numbers with 1 digit, 90 with 2 digits, 900 with 3 digits, etc.

It will not be possible to print all numbers through 999,999,999, however, let us try the number with one fewer digit 9. Consider the following: What is the number of all printed digits of the numbers from 1 through 99,999,999? It is $(1 \times 9) + (2 \times 90) + (3 \times 900) + (4 \times 9,000) + (5 \times 90,000) + (6 \times 900,000) + (7 \times 9,000,000) + (8 \times 90,000,000) = 788,888,889$.

Thus, $1,000,000,000 - 788,888,889 = 211,111,111$ digits remaining. All the further numbers from 100,000,000 onwards have nine digits. Therefore, another

$$\frac{211,111,111}{9} = 23,456,790.111111111111111111111111 \approx 23,456,790$$

numbers can be printed, where the last completely printed number is

$$99,999,999 + 23,456,790 = 100,000,000 + 23,456,789 = 123,456,789.$$

A Numerical Peculiarity

Look at the following pattern and marvel at the relationships:

$$19 \times 1 = 19 \text{ and } 9 + 1 = 10, \text{ and } 1 + 0 = 1$$
$$19 \times 2 = 38 \text{ and } 3 + 8 = 11, \text{ and } 1 + 1 = 2$$
$$19 \times 3 = 57 \text{ and } 5 + 7 = 12, \text{ and } 1 + 2 = 3$$
$$19 \times 4 = 76 \text{ and } 7 + 6 = 13, \text{ and } 1 + 3 = 4$$
$$19 \times 5 = 95 \text{ and } 9 + 5 = 14, \text{ and } 1 + 4 = 5$$
$$19 \times 6 = 114 \text{ and } 11 + 4 = 15, \text{ and } 1 + 5 = 6$$
$$19 \times 7 = 133 \text{ and } 13 + 3 = 16, \text{ and } 1 + 6 = 7$$
$$19 \times 8 = 152 \text{ and } 15 + 2 = 17, \text{ and } 1 + 7 = 8$$
$$19 \times 9 = 171 \text{ and } 17 + 1 = 18, \text{ and } 1 + 8 = 9$$
$$19 \times 10 = 190 \text{ and } 19 + 0 = 19, \text{ and } 1 + 9 = 10$$

Note the end results!

For further amazement, consider the following:

$$9109 \times 1 = 09{,}109 \text{ and } 0+9+1+0+9 = 19$$
$$9109 \times 2 = 18{,}218 \text{ and } 1+8+2+1+8 = 20$$
$$9109 \times 3 = 27{,}327 \text{ and } 2+7+3+2+7 = 21$$
$$9109 \times 4 = 36{,}436 \text{ and } 3+6+4+3+6 = 22$$
$$9109 \times 5 = 45{,}545 \text{ and } 4+5+5+4+5 = 23$$
$$9109 \times 6 = 54{,}654 \text{ and } 5+4+6+5+4 = 24$$
$$9109 \times 7 = 63{,}763 \text{ and } 6+3+7+6+3 = 25$$
$$9109 \times 8 = 72{,}872 \text{ and } 7+2+8+7+2 = 26$$
$$9109 \times 9 = 81{,}981 \text{ and } 8+1+9+8+1 = 27$$

Note the sequence of the resulting numbers!

The Sum of Squares Equals the Sum of More Squares

Here is an arithmetic surprise that sounds very simple and yet requires a little bit of thought. Begin by taking the sum of any three squares and multiplying the sum by 3. We now need to find four squares that will have the same sum. For example, $3(2^2+3^2+4^2) = 87 = 9^2 + 2^2 + 1^2 + 1^2$. Or perhaps as another example: $3(2^2+3^2+3^2) = 66 = 5^2+4^2+4^2+3^2$. This can be sometimes rather frustrating and, yet, also delightfully challenging, which is then considered entertaining.

Explanation: For those who might want to see a justification so that they don't feel that we left them with an unsolvable situation, we provide a simple algebraic proof:

$$3(a^2 + b^2 + c^2) = (a + b + c)^2 + (b^2 - 2bc + c^2)$$
$$+(c^2 - 2ca + a^2) + (a^2 - 2ab + b^2)$$
$$= (a + b + c)^2 + (b - c)^2 + (c - a)^2 + (a - b)^2$$

Using this relationship will also provide other possible solutions.

Sums of Powers

For those who saw the 2015 movie "The Man Who Knew Infinity," they will recall the last scene where the famous Indian mathematician Srinivasa Ramanujan (1887–1920) instantly cites from his hospital bed that the number 1729 is the smallest number that can be expressed as a sum of two cubes in two different ways. That is, $1729 = 12^3 + 1^3 = 10^3 + 9^3$. By the way, 1729 is a number which is divisible by the sum of its digits. That is, $\frac{1729}{1+7+2+9} = \frac{1729}{19} = 91$. From this, we also have another curiosity: $1729 = 19 \times 91$.

Building on the previous curiosity, the number 6578 is the smallest number that can be expressed as a sum of 3 fourth powers in two different ways. That is, $6578 = 1^4 + 2^4 + 9^4 = 3^4 + 7^4 + 8^4$. As a follow-up, one may search for two-digit numbers that can be expressed as a sum of squares in two different ways. One such number is $65 = 8^2 + 1^2 = 7^2 + 4^2$. By the way, the number 65 can also be expressed as the sum of two cubes: $65 = 4^3 + 1^3$. By now, you should be able to see how we can constantly seek such relationships among our numbers. Just searching for such number patterns can be quite amusing and, upon success, rewarding at the same time.

While we are fixed on the sum of squares, we can find a curious arrangement as follows. If we multiply the sum of two different squares by the sum of two other different squares, the result would be the sum of two squares in different ways. We can show this symbolically in the following fashion:

$$(a^2 + b^2) \cdot (c^2 + d^2) = (ac + bd)^2 + (ad - bc)^2 \text{ or}$$
$$(a^2 + b^2) \cdot (c^2 + d^2) = (ac - bd)^2 + (ad + bc)^2$$

Let's see how that works with the numbers $a = 2, b = 5, c = 3$, and $d = 6$ so that

$$(2^2 + 5^2) \cdot (3^2 + 6^2) = 29 \cdot 45 = 1{,}305$$

We then can set up the following: $(2^2 + 5^2) \cdot (3^2 + 6^2) = (2 \cdot 3 + 5 \cdot 6)^2 + (2 \cdot 6 - 3 \cdot 5)^2 = 36^2 + (-3)^2 = 1296 + 9 = 1{,}305$.

And alternatively: $(2^2 + 5^2) \cdot (3^2 + 6^2) = (2 \cdot 3 - 5 \cdot 6)^2 + (2 \cdot 6 + 3 \cdot 5)^2 = 24^2 + 27^2 = 576 + 729 = 1{,}305$.

You might like to try to see if you can come up with another set of two sums of squares.

Guessing the Birthday Date

Alfred says:

"A trickster wanting to impress her audience tells a person that she can guess his birthdate if he does the following arithmetic: Double the number of the days of the month to reach your birthday and then add 5. Multiply this result by 50 and then add the number of the months to reach your birthday month. Then the participant is to tell the trickster the result of the calculations, and she will determine the birthday date."

How and why does that work?

Explanation: Let d be the number of days in the month to reach your birthday ($1 \leq d \leq 31$) and let m be the number of months in the year to reach the birthday month ($1 \leq m \leq 12$).

This can be seen algebraically as $(2d + 5)\,50 + m = 100d + m + 250$.

That indicates that Alfred must simply subtract 250 from the result. The first two digits denote the number of days from the start of the month to reach the birthdate, and the last two digits denote the number of months to reach the birthdate. For example, suppose the calculations is as follows: 18 days to the birthday, doubled and adding 5 yields $2 \times 18 + 5 = 41$, then $41 \times 50 = 2{,}050$, then adding the 10th month (October) leads to a result of 2,060, Alfred then subtracts 250, which yields 1,810, which we then note as 18–10, and that indicates the 18th of October, which was the sought-after birthday.

Natural Numbers and Their Digits

Consider this challenge: How many natural numbers less than 100 million have the property that the first digit (from the left) equals the number of digits?

Explanation: Once again, we will seek a pattern in order to resolve the question. Considering one-digit numbers, this holds only for the number 1. Considering two-digit numbers, this holds for all numbers 2*, where * can be all 10 digits. Now, considering three-digit numbers, we have the numbers of the form 3** of which there are 100 possibilities. The pattern continues so that we have $1 + 10 + 100 + \cdots + 100,000,000 = 111,111,111$ natural numbers.

Determining Remainders of Special Divisions

Begin by considering the following simple question: What is the smallest two-digit number that always yields the remainder 1 when divided by 3, 4, and 5? Now, consider larger numbers, where respective remainders from division by 2, 3, 4, 5, 6, 7, 8, 9, and 10 are 1, 2, 3, 4, 5, 6, 7, 8, and 9, where the remainder is always 1 less than the divisor.

Explanation: We know that the smallest number divisible by 3, 4, and 5 is merely the product of these three numbers. Therefore, by adding 1 to the product, we find the number which divided by these three numbers has a remainder of 1. That is, $3 \cdot 4 \cdot 5 + 1 = 61$. Because the remainders are always 1 less than the divisor, we have to subtract 1 from the respective multiples, and this has to work for all powers of primes which divide the dividend, and this yields

$$\underbrace{2^3 \cdot 3^2 \cdot 5 \cdot 7}_{\substack{\text{divisible by} \\ 2,3,4,5,6,7,8,9,10}} -1 = 2,520 - 1 = 2,519$$

Inverting the Order of the Digits

Let us denote a four-digit number by the digits "*abcd*". Are there four-digit numbers "*abcd*" so that the multiplication by 9 inverts the order of digits: "*abcd*" $\cdot 9 =$ "*dcba*"? If yes, what are these numbers? If no, why not? A further question would be as follows: Are there five-digit numbers "*abcde*" so that the multiplication by 9 inverts the order of digits: "*abcde*" $\cdot 9 =$ "*edcba*"? If yes, what are these numbers? If no, why not?

Explanation: The answer to the first question is yes, as it holds for the number $1089 \cdot 9 = 9801$. We can analyze this as follows: "*abcd*" $\cdot 9 =$ "*dcba*"; this implies that $a = 1$ and $d = 9$ so that "$1bc9$" $\times 9 =$ "$9cb1$"; this yields

$$9000 + 900b + 90c + 81 = 9000 + 100c + 10b + 1, \text{ or}$$

$$90b + 9c + 8 = 10c + b$$

$$89b + 8 = c$$

and from that, we can conclude $b = 0$ and $c = 8$.

Alternative explanation: It is clear that "*abcd*" < 1111. Consider "$1bc9$" $\cdot 9 =$ "$9cb1$", where for the value of b, there are only two possibilities $b = 0$, $b = 1$: In case of $b = 1$, we have "$11c9$" $\cdot 9 =$ "$9c11$", but that is impossible because on the one hand, this would yield $c = 9$ ($100 \cdot 9 = 900$), and on the other hand, the number would be greater than 1111. Hence, $b = 0$ so that "$10c9$" $\cdot 9 =$ "$9c01$". That means $c \cdot 9$ (tenths place) must end with 2 because at the ones place, we have $9 \cdot 9 = 81$, carry over 8, and at the tenths place, we must have 0; thus, only $c = 8$ is possible.

This can be extended to a five-digit number, where $10989 \cdot 9 = 98901$, where symbolically we have "*abcde*" $\cdot 9 =$ "*edcba*", which yields $a = 1$ and $e = 9$ so that "$1bcd9$" $\cdot 9 =$ "$9dcb1$". This leads us to the following:

$$90000 + 9000b + 900c + 90d + 81 = 90000 + 1000d + 100c + 10b + 1$$
$$\Rightarrow 899b + 80c + 8 = \underbrace{91d}_{\leq 819}$$

Therefore, $b = 0$ and $80c + 8 = 91d$. Hence, $91d$ must end with $8 \Rightarrow d = 8$, $c = 9$.

Another alternative explanation: It is clear that "*abcde*" < 11111, hence in "$1bcd9$" $\cdot 9 =$ "$9dcb1$", where for the value for b there are only two possibilities $b = 0$, $b = 1$. In case of $b = 1$, we had "$11cd9$" $\cdot 9 =$ "$9dc11$", but that is impossible because then we would have $d = 9$ ($1000 \cdot 9 = 9000$): "$11c99$" $\cdot 9 =$ "$99c11$" but that contradicts "*abcde*" < 11111. Hence, $b = 0$: "$10cd9$" $\cdot 9 =$ "$9dc01$", which means $d \cdot 9$ (tenths place) must end with 2 because in the ones place, we have $9 \cdot 9 = 81$, carry over 8, and at the tenths place, we finally must have 0, thus, only $d = 8$ is possible: "$10c89$" $\cdot 9 =$ "$98c01$" $\Rightarrow$ "$c89$" $\cdot 9 =$ "$8c01$" $\Rightarrow c = 9$.

Palindromic Numbers

There are certain categories of numbers that have particularly strange characteristics. Here, we consider numbers that read the same in both directions: left to right, and right to left. These are called *palindromic numbers*. First, note that a palindrome can also be a word, phrase, or sentence that reads the same in both directions. Figure 2.1 shows a few amusing palindromes.

A

EVE

CIVIC

LEVEL

RADAR

REVIVER

RACECAR

ROTATOR

DON'T NOD

LEPERS REPEL

MADAM I'M ADAM

STEP NOT ON PETS

NEVER OD OR EVEN

PULL UP IF I PULL UP

NO LEMONS, NO MELON

DENNIS AND EDNA SINNED

ABLE WAS I ERE I SAW ELBA

WAS IT A CAR OR A CAT I SAW

A MAN, A PLAN, A CANAL, PANAMA

A SANTA LIVED AS A DEVIL AT NASA

SUMS ARE NOT SET AS A TEST ON ERASMUS

ANNE, I VOTE MORE CARS RACE ROME TO VIENNA

ON A CLOVER, IF ALIVE, ERUPTS A VAST, PURE EVIL; A FIRE VOLCANO

DOC, NOTE I DISSENT:A FAST NEVER PREVENTS A FATNESS. I DIET ON COD

Figure 2.1

A palindrome in mathematics would be a number such as 666 or 123321 that reads the same in either direction. For example, the first five powers of 11 are palindromic numbers:

$$11^0 = 1$$
$$11^1 = 11$$
$$11^2 = 121$$
$$11^3 = 1331$$
$$11^4 = 14641$$

Once again, using a calculator, there are some unusual aspects, which result from squaring numbers consisting of all 1s, which are often called *repunits*. Unexpectedly, these powers of repunits result in palindromic numbers, as we can see here:

$$11^2 = 121$$
$$111^2 = 1331$$
$$1111^2 = 1234321$$
$$11111^2 = 123454321$$
$$111111^2 = 12345654321$$
$$1111111^2 = 1234567654321$$
$$11111111^2 = 123456787654321$$
$$111111111^2 = 12345678987654321$$

A small curiosity is that the second smallest number with an even number of digits whose square is a palindrome with an even number of digits is the number 798,644 since $798,644^2 = 637,832,238,736$, which is a palindrome.

Now for the entertaining aspect of palindromic numbers. Here, we have a procedure to see how a palindromic number can be generated from a given number. All you need to do is to continually add a given number to its reversal (that is, the number written in the reverse order of digits) until you arrive at a palindrome. For example, a palindrome can be reached with a single addition such as with the starting number 23: the sum $23 + 32 = 55$, a palindrome.

Or it might take two steps, such as with the starting number 75: The two successive sums are $75 + 57 = 132$ and $132 + 231 = 363$, which lead us to a palindrome. Or it might take three steps, such as with the starting number 86: $86 + 68 = 154$, $154 + 451 = 605$, and $605 + 506 = 1111$.

The starting number 97 will require 6 steps to reach a palindrome; the starting number 98 will require 24 steps to reach a palindrome. It is important to be cautioned about using the starting number 196 as this one has not yet been shown to produce a palindromic number, even when tried with over three million reversal additions. We still do not know if this number will ever reach a palindrome.

There are a few quirky results in this procedure. If you *were* to try to apply this procedure to 196, you would eventually — at the 16th addition — reach the number 227574622. Yet, amazingly, you would also reach that same sum at the 15th step of the attempt to get a palindrome from the starting number 788. This would then tell you that applying the procedure to the number 788 has also not yet been shown to reach a palindrome. As a matter of fact, among the first 100,000 natural numbers, there are 5,996 numbers for which we have not yet been able to show that the procedure of reversal additions will lead to a palindrome. Some of the starting numbers of these non-palindrome results are : 196, 691, 788, 887, 1675, 5761, 6347, and 7436.

You might want to take this to another level by showing some unusual aspects in this process. For example, using this procedure of reversal and addition, we find that some numbers yield the same palindrome in the same number of steps, such as 554, 752, and 653, which all produce the palindrome 11011 in three steps. In general, all integers, in which the corresponding digit pairs symmetric to the middle 5 have the same sum, will produce the same palindrome in the same number of steps. The three sample numbers, 554, 752, and 653, have this characteristic since the pair of digits on either side of the middle 5 have the same sum, namely, 9.

There are other integers that produce the same palindrome, yet with a different number of steps, such as the number 198, which, with repeated reversals and additions, will reach the palindrome 79497 in five steps, while the number 7299 will reach this same number in two steps, that is, 7299 + 9927 = 79497.

We can determine the number of additions that we will have to do to reach a palindrome using this procedure. For a two-digit number ab with digits $a \neq b$, the sum $a + b$ of its digits determines the number of steps needed to produce a palindrome. Clearly, if the sum of the digits is less than 10, then only one step will be required to reach a palindrome, for example, $25 + 52 = 77$. If the sum of the digits is 10, then $ab + ba = 110$ and $110 + 011 = 121$, so two steps will be required to reach the palindrome. The number of steps required for each of the two-digit sums 11, 12, 13, 14, 15, 16, and 17 to reach a palindromic number is 1, 2, 2, 3, 4, 6, and 24, respectively.

Now, we can take this yet to another level to appreciate some unusual aspects of palindromic numbers. We can arrive at some lovely patterns when dealing with palindromic numbers. For example, some palindromic numbers when squared also yield a palindrome. For example, $22^2 = 484$ and $212^2 = 44944$. On the other hand, there are also some palindromic numbers that, when squared, do not yield a palindromic number, such as $545^2 = 297,025$. Of course, there are also non-palindromic numbers that, when squared, yield a palindromic number, such as $26^2 = 676$ and $836^2 = 698,896$. These are just some of the entertainments that numbers provide. A motivated reader may want to search for other such curiosities.

Taking Palindromic Numbers Further

There are also some palindromic numbers that, when cubed, again yield palindromic numbers.

To this group belong all numbers of the form $n = 10^k + 1$, for $k = 1, 2, 3, \ldots$. When n is cubed, it yields a palindromic number, which has $k - 1$ zeros between each consecutive pair of 1,3,3,1, as we can see with the following examples:

$k = 1, n = 11$	$11^3 = 1331$
$k = 2, n = 101$	$101^3 = 1030301$
$k = 3, n = 1001$	$1001^3 = 1003003001$
$k = 7, n = 10000001$	$10000001^3 = 1000000300000030000001$

Using algebra, one can show that this pattern continues:

$$\left(10^k + 1\right)^3 = 10^{3k} + 3 \cdot 10^{2k} + 3 \cdot 10^k + 1 \text{ and written in another way,}$$

$$1\underbrace{0 \ldots 01}_{k}{}^3 = 1\overbrace{\underbrace{0 \ldots 03}_{k}\underbrace{\underbrace{0 \ldots 03}_{k}\underbrace{0 \ldots 01}_{k}}_{2k}}^{3k}.$$

Especially in this representation, one can see the pattern immediately: As the value of k increases by 1, the number of zeros in each block increases by 1!

We can continue to generalize to reach some interesting patterns, such as when n consists of three 1s and any even number of zeros symmetrically placed between the end 1s when cubed will give us a palindrome, such as,

$$111^3 = 1367631$$
$$10101^3 = 1030607060301$$
$$1001001^3 = 1003006007006003001$$
$$100010001^3 = 1000300060007000600030001$$

Taking this even another step further, we find that when n consists of four 1s and 0s in a palindromic arrangement, where the places between the 1s do not have the same number of 0s, then n^3 will also be a palindrome, as we can see with the following examples:

$$11011^3 = 1334996994331$$
$$10100101^3 = 1030331909339091330301$$
$$10010001001^3 = 1,003,003,301,900,930,390,091,033,003,001$$

However, when the same number of zeros appears between the 1s, then the cube of the number will not result in a palindrome, as in the following example: $1010101^3 = 1030610121210060301$. As a matter of fact, the number 2201 is the only non-palindromic number, which is less than $280,000,000,000,000$, and, when cubed, yields a palindrome $2201^3 = 10662526601$.

Just for further amusement, consider the following pattern with palindromic numbers:

$$12321 = \frac{333 \cdot 333}{1+2+3+2+1}$$

$$1234321 = \frac{4444 \cdot 4444}{1+2+3+4+3+2+1}$$

$$123454321 = \frac{55555 \cdot 55555}{1+2+3+4+5+4+3+2+1}$$

$$12345654321 = \frac{666666 \cdot 666666}{1+2+3+4+5+6+5+4+3+2+1}$$

and so on until the digit 9 is in the middle.

An ambitious reader may search for other patterns involving palindromic numbers.

The Common Divisor of Palindromic Numbers

Now that we are thoroughly familiar with palindromic numbers, we can consider the following question: What is the greatest common divisor of all four-digit palindrome numbers? The answer is the number 11. Now, let's see why that is the case.

Explanation: A four-digit palindrome is of the form *"anna"* $= a1001 + n110 = (a \times 7 \times 11 \times 13) + (n \times 2 \times 5 \times 11)$ with the possibilities of $a \in \{1, \ldots, 9\}$ and $n \in \{0, \ldots, 9\}$, where also $a = n$ is possible.

A Surprising Pattern of Odd Numbers

Odd numbers can be presented in such a way that they generate cubic numbers. This can be quite surprising as well as enchanting.

$$1 = \qquad\qquad 1 = 1^3$$
$$3 + 5 = \qquad\qquad 8 = 2^3$$
$$7 + 9 + 11 = \qquad\qquad 27 = 3^3$$
$$13 + 15 + 17 + 19 = \qquad\qquad 64 = 4^3$$
$$21 + 23 + 25 + 27 + 29 = \qquad\qquad 125 = 5^3$$
$$31 + 33 + 35 + 37 + 39 + 41 = \qquad\qquad 216 = 6^3$$

Consider what the sum of the tenth line will be. The tenth line would have a sum equal to $10^3 = 10{,}000$. Amazement is typically generated by how such a simple relationship can generate cubes.

The Digits Remain in Use

Simple multiplication (of course, having a calculator handy makes the experience much more less cumbersome) can generate some unexpected results. Here are several multiplication examples that continue to use only the digits of the two given numbers in the product; as you will note, in the first product, only the digits 0, 1, 3, and 5 are used.

$$30 \times 51 = 1{,}530$$
$$21 \times 87 = 1{,}827$$
$$80 \times 86 = 6{,}880$$
$$60 \times 21 = 1{,}260$$
$$93 \times 15 = 1{,}395$$
$$41 \times 35 = 1{,}435$$

Naturally, this can be taken for larger numbers as well. For example, take a pair of larger numbers and consider a case in which each of the digits in the original multiplication is used in the product as many times as it appears in the original modification.

$$9{,}162{,}361{,}086 \times 1{,}234{,}554{,}321 = 11{,}311{,}432{,}469{,}283{,}552{,}606$$

Some More Cute Number Patterns

It is not very difficult to note that one can express a number as the sum of three other numbers. However, with the number 118, we can also express it as the sum of four arrangements of three numbers, and the amazing thing is that the product of each of these sets of three numbers is the same for all four groups of three, namely, 37,800. Take a look here:

$$15 + 40 + 63 = 118 \text{ and } 15 \times 40 \times 63 = 37{,}800$$
$$14 + 50 + 54 = 118 \text{ and } 14 \times 50 \times 54 = 37{,}800$$
$$21 + 25 + 72 = 118 \text{ and } 21 \times 25 \times 72 = 37{,}800$$
$$18 + 30 + 70 + 118 \text{ and } 18 \times 30 \times 70 = 37{,}800$$

More amazingly, this number 118 is the smallest number for which this can be done. You might want to challenge yourself to come up with some other such arrangements for other numbers.

Here is another nice number relationship to enjoy! Admire the symmetry:

$$13^3 - 3^7 = 2197 - 2187 = 13 - 3$$
$$5^3 - 2^7 = 125 - 128 = -(5 - 2)$$

Do not try to find another such a pair of numbers because no other such pair has yet been found!

There are many other curious number relationships. We will present a few of them here. Note that in each case, the exponents are consecutive:

$$43 = 4^2 + 3^3$$
$$63 = 6^2 + 3^3$$
$$135 = 1^1 + 3^2 + 5^3$$
$$175 = 1^1 + 7^2 + 5^3$$
$$518 = 5^1 + 1^2 + 8^3$$
$$598 = 5^1 + 9^2 + 8^3$$
$$1306 = 1^1 + 3^2 + 0^3 + 6^4$$
$$1676 = 1^1 + 6^2 + 7^3 + 6^4$$
$$2427 = 2^1 + 4^2 + 2^3 + 7^4$$

And here is one where the exponents match the base:

$$3435 = 3^3 + 4^4 + 3^3 + 5^5$$

You might want to see if there are other such relationships or even some such as

$$244 = 1^3 + 3^3 + 6^3, \text{ and } 136 = 2^3 + 4^3 + 4^3$$

Note the amazing relationship here!

If you want to keep this to one exponential number, you can do this with the following numbers:

$$153 = 1^3 + 5^3 + 3^3$$
$$370 = 3^3 + 7^3 + 0^3$$
$$371 = 3^3 + 7^3 + 1^3$$
$$407 = 4^3 + 0^3 + 7^3$$

We can take this even a step further by considering the four-digit numbers, such as the following:

$$1634 = 1^4 + 6^4 + 3^4 + 4^4$$
$$8208 = 8^4 + 2^4 + 0^4 + 8^4$$
$$9474 = 9^4 + 4^4 + 7^4 + 4^4$$

A Surprising Pattern

Begin by considering the following problem, which on the surface seems rather harmless but could get a bit cumbersome. What is the sum of $1^3 + 2^3 + 3^3 + 4^3 + \cdots + 9^3 + 10^3$?

If carefully done with the aid of a calculator, this should yield the correct answer. However, if we do not have a calculator at hand, the multiplication and addition could prove quite complicated and messy! Let's see how we might solve the problem by searching for a pattern. Let's organize our data as shown below:

1^3	$= (1)$	$= 1$	$= 1^2$
$1^3 + 2^3$	$= (1 + 8)$	$= 9$	$= 3^2$
$1^3 + 2^3 + 3^3$	$= (1 + 8 + 27)$	$= 36$	$= 6^2$
$1^3 + 2^3 + 3^3 + 4^3$	$= (1 + 8 + 27 + 64)$	$= 100$	$= 10^2$

Note that the number bases in the final column (namely, 1, 3, 6, 10, ...) are *triangular numbers,* which are numbers that represent points that can be arranged to form equilateral triangles; they are 1, 3, 6, 10, 15, 21, ... The nth triangular number is formed by taking the sum of the first n integers. That is, the first triangular number is 1. The second triangular number is $3 = (1 + 2)$. The third triangular number is $6 = (1 + 2 + 3)$. The fourth triangular number is $10 = (1 + 2 + 3 + 4)$, and so on.

Thus, we can rewrite our problem as follows:

1^3	$= (1)^2$	$= 1^2 = 1$
$1^3 + 2^3$	$= (1 + 2)^2$	$= 3^2 = 9$
$1^3 + 2^3 + 3^3$	$= (1 + 2 + 3)^2$	$= 6^2 = 36$
$1^3 + 2^3 + 3^3 + \cdots + 9^3 + 10^3$	$= (1 + 2 + 3 + \cdots + 9 + 10)^2$	$= 55^2 = 3025$

By this point, you should have gotten a "feel" for the advantage of looking for a pattern in solving a problem. It may take some effort to find a pattern, but when one is discovered, it not only simplifies the problem greatly but also once again demonstrates the beauty of mathematics.

Equal Number of Numbers with Equal Sums

The challenge is to determine for which numbers n can the set $\{1, 2, \ldots, n\}$ be partitioned into two subsets with an equal number of numbers and equal sums.

For example, it is impossible with $n = 3$, having the set $\{1, 2, 3\}$, yet it is possible for $n = 4$, where we have $\{1, 2, 3, 4\}$, which can be partitioned as $\{1, 4\}$ and $\{2, 3\}$, where in both cases 2 numbers, with equal sums of 5.

How might the challenge be modified if we are not limited to an equal product of numbers and only have equal sums? In that case, we would have $n = 3$ as a solution since $\{1, 2, 3\}$ can be partitioned in $\{1, 2\}$ and $\{3\}$, where in both cases the sum equals 3. Before reading the explanation, a enthusiastic reader might seek other such examples as shown above.

Explanation: We can show that it is possible for $n = 4, 8, 12, \ldots$, so that $n = 4k$. For example, when $n = 8$, we have two possibilities $\{3, 4, 5, 6\}$ and $\{1, 2, 7, 8\}$ with sums of 18, as well as $\{2, 3, 6, 7\}$ and $\{1, 4, 5, 8\}$ with sums of 18, as we shown in Figure 2.2.

When n is a multiple of 4 ($n = 4k$), it works with pairs of equal sums from the "boundaries" to the "center." One of the possibilities is shown in

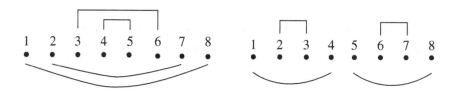

Figure 2.2

Figure 2.2. The odd numbers such as $(4k + 1$ and $4k + 3)$ for n cannot work because one cannot divide $\{1, 2, \ldots, n\}$ into two subsets with equal number of numbers. Furthermore, when $n = 4k + 2$, it is also not possible because there the sum $1 + 2 + \cdots + n$ would be odd, and one could not get two equal sums. Although it is good to know it, one does not necessarily need the formula for the sum of this arithmetic series: $1 + 2 + \cdots + n = \frac{n(n+1)}{2}$. The sum $1 + 2 + 3 + 4 = 10$ is even; with the series up to 5, the sum is odd, and with the series up to 6, the sum is again odd. Yet, when the series extends to 7, the sum is even, and when the series extends to 8, the sum is also even; this "pattern," of course, continues beyond 8, hence, one can conclude that the sum from 1 to 6 is odd (21), as the sums are also odd for series up to 10, 14, 18, ..., $4k + 2$, ...

We seek to have the sum $1 + 2 + \cdots + n$ to be even. Such a sum is even for the first time with $n = 3$, where $1 + 2 + 3 = 6$, and again with $n = 4$, where $1 + 2 + 3 + 4 = 10$; yet with $n = 5$, the sum is odd, and it is again odd with $n = 6$; then again, the sum is even with $n = 7$ and $n = 8$; then again, twice odd, etc. That means all the multiples of 4 are a possibility (as shown above), but now their predecessors are also possible, namely, $n = 3, 7, 11, \ldots n = 4k - 1$. If one has a partition in two subsets with equal sums for n, one has such a partition also for $n + 4$ (see ·Figure 2.3).

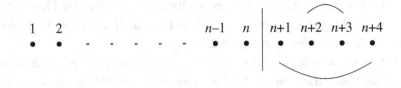

Figure 2.3

And since we have a concrete partition for $n = 3$, we know that altogether there are such partitions for the numbers, $n = 4k$ and $n = 4k - 1$ $(k = 1, 2, 3, \ldots)$.

Splitting Numbers

We can even split numbers other than into individual digits and still end up with some spectacular results such as the following:

$$1,233 = 12^2 + 33^2$$

$$8,833 = 88^2 + 33^2$$

$$5,882,353 = 588^2 + 2353^2$$

$$94,122,353 = 9412^2 + 2353^2$$

$$1,765,038,125 = 17650^2 + 38125^2$$

$$2,584,043,776 = 25840^2 + 43776^2$$

There are many more such amazements as we focus on a "reverse" situation, where we could take the difference of the squares of a split number rather than the sum of squares. For instance, we reverse the two parts so that from 48, we would consider the reverse and split the digits 8 and 4 and subtract the squares: $48 = 8^2 - 4^2$. Here are several more such examples:

3,468 to be split as 34 and 68 so that $68^2 - 34^2 = 3,468$

16,128 to be split as 16 and 128 so that $128^2 - 16^2 = 16,128$

34,188 to be split as 34 and 188 so that $188^2 - 34^2 = 34,188$

216,513 to be split as 216 and 513 so that $513^2 - 216^2 = 216,513$

416,768 to be split as 416 and 768 so that $768^2 - 416^2 = 416,768$

2,661,653 to be split as 266 and 1653 so that $1653^2 - 266^2 = 2,661,653$

59,809,776 to be split as 5980 and 9776 so that $9776^2 - 5980^2 = 59,809,776$

There are many more such examples that you might feel compelled to search for.

We can then take this a step further in our effort to further highlight unusual relationships. Consider displaying portions of numbers as the sum of cubes, as we show with a few examples here: 41,833, which we can split up as follows to get the sum of cubes: $4^3 + 18^3 + 33^3 = 41{,}833$.

$$221{,}859 = 22^3 + 18^3 + 59^3$$

$$444{,}664 = 44^3 + 46^3 + 64^3$$

$$487{,}215 = 48^3 + 72^3 = 15^3$$

$$336{,}701 = 33^3 + 67^3 + 01^3$$

$$982{,}827 = 98^3 = 28^3 = 27^3$$

$$983{,}221 = 98^3 = 32^3 + 21^3$$

$$166{,}500{,}333 = 166^3 + 500^3 + 333^3$$

Once again, this is not an exhaustive list and there are more such numbers that lend themselves to this unusual splitting arrangement.

We can always look for nice relationships between numbers. With some creativity, we can establish another form of "friendliness" between numbers. Some of them can be truly mind-boggling! Take, for example, the pair of numbers 6,205 and 3,869.

At first glance, there seems to be no apparent relationship. But with some luck and imagination, we can get some fantastic results. $6{,}205 = 38^2 + 69^2$ and $3{,}869 = 62^2 + 05^2$.

We can even find another pair of numbers with a similar relationship. Consider these: $5{,}965 = 77^2 + 06^2$ and $7{,}706 = 59^2 + 65^2$. Quite spectacular!

A Random Division

David writes the 16 digits 2, 2, 3, 3, 4, 4, 5, 5, 6, 6, 7, 7, 8, 8, 9, 9 in an arbitrary order and randomly inserts the division symbol "÷" between two digits indicating a division. Is it possible that the result of this division is exactly 2? Hint: Use the divisibility rule for 3, where the sum of the digits must be divisible by 3, which was presented on page 13.

Explanation: Let G be the written number 2,233,445,566,778,899 without including the "÷" sign in any place. Then, by using the well-known rule for divisibility by 3, namely, that a number is divisible by 3 if, and only if, the sum of its digits is divisible by 3, we know that G is not divisible by 3 since the sum of the digits of the number is 88, which is not divisible by 3. Therefore, no divisor of G can be a multiple of 3 (later, this fact will serve as a contradiction, see the following). Now, let us denote the left part of G (i.e., to the left of the "÷" sign) with L, and denote the right part of G with R (i.e., to the right of the "÷" sign). For an indirect proof, let us assume that the division $L \div R = 2$ was without a remainder, then it would follow directly from the algorithm of division that the result of $G \div R$

is $G \div R = \overbrace{\underbrace{* \cdots *}_{L \atop m \text{ digits}} \underbrace{* \cdots *}_{R \atop n \text{ digits}}}^{G} \div R = 2\underbrace{0 \cdots 0 1}_{n \text{ digits}}$, where the number n of

digits after 2 in the result (quotient) equals the number of digits in R.

We can also explain this phenomenon in another way; we will keep the explanation simple and not use all 16 digits as originally proposed. Consider, for example, $L = 482$ and $R = 241$, then $L \div R = \frac{482}{241} = 2$. Now, consider $L \div R = \frac{482241}{241} = 2001$. This could also be written as $\frac{G}{R} = \frac{2 \cdot 241 \cdot 10^3 + 241}{241} = 2 \cdot 1000 + 1 = 2001$. Again, we see that the three digits after the digit 2 in 2001 come from the three digits in 241.

Now, we have the desired contradiction: From the above, we know that G and no divisor of G can be divisible by 3. But the assumption $L \div R = 2$ led us to $G \div R = 2\underbrace{0 \ldots 0 1}_{n \text{ digits}}$ which is a divisor of G divisible by 3, which

is a contradiction.

The Surprising Sum and Product of Special Fractions

Given three positive (randomly selected) integers, such as 4, 5, and 14, we shall create three fractions using these numbers in the following fashion: The numerator of each fraction will be the sum of two of these numbers and the denominator will be the third number so that we get $\frac{4+5}{14}, \frac{4+14}{5}, \frac{5+14}{4} = \frac{9}{14}, \frac{18}{5}, \frac{19}{4}$. Next, we will subtract the sum of these three fractions $\frac{9}{14} + \frac{18}{5} + \frac{19}{4} = \frac{90+504+665}{140} = \frac{1259}{140} = \frac{2518}{280}$ from their product $\frac{9}{14} \cdot \frac{18}{5} \cdot \frac{19}{4} = \frac{3078}{280}$ to get $\frac{3078}{280} - \frac{2518}{280} = \frac{560}{280} = 2$. You may want to take three other starting numbers and follow this procedure. Amazingly, you will find that the final result will always be 2.

Explanation: To justify that the final result will always be 2, we would use simple algebra. Let a, b, c be our starting integers:

$$\frac{a+b}{c} \cdot \frac{b+c}{a} \cdot \frac{a+c}{b} - \left(\frac{a+b}{c} + \frac{b+c}{a} + \frac{a+c}{b}\right)$$

$$= \frac{(ab+ac+b^2+bc) \cdot (a+c)}{abc} -$$

$$\frac{ab \cdot (a+b) + bc \cdot (b+c) + ac \cdot (a+c)}{abc}$$

$$= \frac{a^2b\,a^2c + ab^2 + abc + abc + ac^2 + b^2c + bc^2}{abc} -$$

$$\frac{a^2b + ab^2 + b^2c + bc^2 + a^2c + ac^2}{abc}$$

$$= \frac{2abc}{abc} = 2$$

Friendly Numbers

Within the realm of numerical novelties, there are numbers that are considered to be "friendly numbers." Mathematicians have decided that two numbers are to be considered friendly (or as sometimes used in the more sophisticated literature, "amicable") if the sum of the proper divisors[1] (or factors) of one number equals the second number *and* the sum of the proper divisors of the second number equals the first number as well. Just take a look at the smallest pair of friendly numbers: 220 and 284. The divisors (or factors) of **220** are 1, 2, 4, 5, 10, 11, 20, 22, 44, 55, and 110. Their sum is $1 + 2 + 4 + 5 + 10 + 11 + 20 + 22 + 44 + 55 + 110 = \mathbf{284}$.

The divisors of **284** are 1, 2, 4, 71, and 142, and their sum is $1 + 2 + 4 + 71 + 142 = \mathbf{220}$.

This shows that these two numbers can be considered *friendly numbers*.

---

[1] Proper divisors are all the divisors, or factors, of the number except the number itself. For example, the proper divisors of 6 are 1, 2, and 3 but not 6.

A second pair of friendly numbers, which were discovered by the famous French mathematician Pierre Fermat (1601–1665), is 17,296 and 18,416. In order for us to establish their friendliness relationship, we need to find all of their prime factors, which are $17,296 = 2^4 \times 23 \times 47$ and $18,416 = 2^4 \times 1151$. Then, we need to create all the numbers from these prime factors as follows:

The sum of the factors of 17,296 is $1+2+4+8+16+23+46+47+92+94+184+188+368+376+752+1081+2162+4324+8648 = \underline{18,416}$.

The sum of the factors of 18,416 is $1 + 2 + 4 + 8 + 16 + 1151 + 2302 + 4604 + 9208 = \underline{17,296}$.

Once again, we note that the sum of the factors of 17,296 is equal to 18,416, and conversely, the sum of the factors of 18,416 is equal to 17,296. This qualifies them to be considered a pair of friendly numbers.

There are many more such pairs. Here are a few more pairs of friendly numbers:

> 1,184 and 1,210
>
> 2,620 and 2,924
>
> 5,020 and 5,564
>
> 6,232 and 6,368
>
> 10,744 and 10,856
>
> 9,363,584 and 9,437,056
>
> 111,448,537,712 and 118,853,793,424

An ambitious reader might want to verify the above pairs' "friendliness!"

Explanation: For the experts, the following is one method for finding pairs of friendly numbers: Let $a = 3 \cdot 2^n - 1, b = 3 \cdot 2^{n-1} - 1,$ and $c = 3^2 \cdot 2^{2n-1} - 1$, where n is an integer greater than or equal to 2, and a, b, and c are all prime numbers. It then follows that $2^n ab$ and $2^n c$ are friendly numbers. We should note that for $n = 2, 4$, and 7, we have a, b, and c that are all prime for n less than or equal to 200. Another form of friendliness can be seen with the following examples:

$$3869 = 62^2 + 05^2 = 6205 = 38^2 + 68^2$$
$$5965 = 77^2 + 06^2 = 7706 = 59^2 + 65^2$$

Are there other numbers that exhibit such friendliness?

We can even set up an analogous cycle using cubes: Starting with 55: $5^3 + 5^3 = 250$, then 250: $2^3 + 5^3 + 0^3 = 133$, and then 133: $1^3 + 3^3 + 3^3 = 55$, which is the number we started with. This can be done with other sequences of numbers, such as

$$136, 244, 136$$

$$919, 1459, 919$$

$$160, 217, 352, 160$$

The Magic of Square Numbers

Let's consider a certain "magic" of square numbers. But first, let's take a slight detour to marvel at another curiosity. Sometimes, peculiarities are so simple and yet can be interesting. Take, for example, the fact that there are only two numbers, 2 and 11, where their squares increased by 4 will yield a cube.

$2^2 = 4$, then by adding 4, we get $4 + 4 = 8 = 2^3$

$11^2 = 121$, then by adding 4, we get $121 + 4 = 125 = 5^3$

Now, let's take a look at a list of square natural numbers and see if there is any pattern to be recognized. Patterns always seem to provide enrichment or enlightenment.

1^2	2^2	3^2	4^2	5^2	6^2	7^2	8^2	9^2	10^2	11^2
1	4	9	16	25	36	49	64	81	100	121
12^2	13^2	14^2	15^2	16^2	17^2	18^2	19^2	20^2	21^2	
144	169	196	225	256	289	324	361	400	441	

One thing that may be quickly noted among the square numbers listed is that the units digits, which we have bold underlined above, follow a specific pattern, namely, 1, 4, 9, 6, 5, 6, 9, 4, 1, 0, **1, 4, 9, 6, 5, 6, 9, 4, 1, 0,** 1,.... This pattern will continue without end. Seeing this, one would be able to surmise that there are certain digits, which can never appear in the units-digit position of a square number since they are missing from the repetitions list. That is, the digits 2, 3, 7, and 8 will never be the units digit of a square number. Furthermore, these numbers separated by the zero is a palindromic arrangement which can easily be spotted in the sequence

1, 4, 9, 6, 5, 6, 9, 4, 1, 0, **1, 4, 9, 6, 5, 6, 9, 4, 1, 0,** 1...
and will continue ongoing.

There is probably no limit to the number of curiosities we can offer about square numbers. For example, the numbers 13 and 31, which are reversals of one another, have, respectively, squares that are also reversals of one another, that is, 169 in 961. Furthermore, if we take the product of these two numbers, we get $169 \times 961 = 162{,}409 = 403^2$; yet another square appears. If we want to take this a step further, the sum of the digits of 169 is $1 + 6 + 9 = 16 = 4^2$, and the sum of the digits of the square root of 169, which is 13, is $1 + 3 = 4 = 2^2$, are in both cases square numbers. To add to this surprisingly beautiful relationship, there is another pair of numbers that has the same characteristic: These numbers are 12 and 21. If we follow the same pattern as we did with the numbers 13 and 31, we will get $12^2 = 144$ and $21^2 = 441$. The product of these two numbers is $144 \times 441 = 63{,}504 = 252^2$. In addition, $1 + 4 + 4 = 9 = 3^2$ and $1 + 2 = 3$, which is analogous to what we have done with the numbers 13 and 31.

While we are admiring square numbers, there are numbers, called *automorphic numbers*, whose squares end in the same digits, such as

$$5^2 = 2\mathbf{5}$$
$$6^2 = 3\mathbf{6}$$
$$76^2 = 5{,}7\mathbf{76}$$
$$376^2 = 141{,}\mathbf{376}$$
$$625^2 = 390{,}\mathbf{625}$$
$$90{,}625^2 = 8{,}212{,}\mathbf{890{,}625}$$
$$890{,}625^2 = 793{,}212{,}\mathbf{890{,}625}$$
$$1{,}787{,}109{,}376^2 = 3{,}193{,}759{,}921{,}\mathbf{787{,}109{,}376}$$
$$8{,}212{,}890{,}925^2 = 67{,}451{,}572{,}418{,}\mathbf{212{,}890{,}625}$$

After observing this pattern, the question is: How can we create other automorphic numbers?

Explanation: Suppose we take the next-to-last automorphic number above and chop off several of its left-side digits so that we consider the number **921,787,109,376,** which when we square it, we get the number 849,691,475,011,76**1,787,109,376**. You will note that the last 10 digits are the same. This can be done with any of the above automorphic numbers, as we can also see with the previously calculated numbers: $90,625^2 =$ 8,212,8**90,625** and $890,625^2 = 793,212,$**890,625**.

At this point, you may like to experiment by tagging on to the front of some of these numbers a few random digits keeping the terminal digits as shown above and finding that in each case, automorphic numbers will be created. Keep in mind that at most, there are two suffix groups of a specific number of digits that can be used to create automorphic numbers. For example, the numbers 625 and 376 are the only numbers of three digits that can be used to make automorphic numbers, such as the number $1,234,$**625**$^2 =$ 1,524,298,890,**625**.

We should note that the number 90,625 is the only five-digit automorphic number. We can see a few uses of that five-digit automorphic number above.

Here are the automorphic numbers up to 10^{15}:

1, 5, 6, 25, 76, 376, 625, 9376, 90625, 109376, 890625, 2890625, 7109376, 12890625, 87109376, 212890625, 787109376, 1787109376, 8212890625, 18212890625, 81787109376, 918212890625, 9918212890625, 40081787109376, 59918212890625, 259918212890625, 740081787109376

At this point, you have lots to experiment with to try to create numbers whose square ends up with the same end digits as the original number. Lots of fun lurks in the future!

What Are Lucky Days?

A day should be called *lucky day* if, when written in the format MM/DD/YYYY, it uses eight different digits (days and months with numbers less than 10 are preceded by a 0). For example, the date 04/23/1965 was a lucky day. We are now in the 21st century. When is the next lucky day after 2023?

Explanation: The next lucky day is 06/17/2345. In all coming years of the form 20**, there cannot be a lucky day because one cannot find a corresponding month since we cannot use the 0 which covers the first 10 months, and we cannot use November or December as well. In the years 21**, the month must contain 0, but then it is impossible to find a corresponding day. Next, we are looking at the years of the form 23**. The month of a lucky day then must contain 0 and hence the day must start with 1. The next possible year is thus 2345 with the first possible month 06 and the first possible day 17. Therefore, the next lucky day is June 17, 2345.

Typographical Errors That *Are* Correct

Just for entertainment, here are some typographical errors that turn out to be correct. Consider the following, where the × was missing and misplaced, such as with

$$73 \times 9 \times 42 = 7 \times 3942$$

$$73 \times 9 \times 420 = 7 \times 39420$$

Then there are some where the × and the exponents dropped out, such as with

$$2^5 \times \frac{25}{31} = 25\frac{25}{31}$$

$$2^5 \times 9^2 = 2592$$

$$3^4 \times 425 = 34425$$

$$3^4 \times 4250 = 344250$$

$$11^2 \times 9\frac{1}{3} = 1129\frac{1}{3}$$

$$21^2 \times 4\frac{9}{11} = 2124\frac{9}{11}$$

$$13^2 \times 7\frac{6}{7} = 1327\frac{6}{7}$$

There are also some very complicated versions of these curious typographical errors, such as

$$13^2 \times 7857142\frac{6}{7} = 1327857142\frac{6}{7}.$$

There are many more such strange coincidences, where symbols are deleted or misread and still the correct answer results. We provide this merely as an entertaining activity!

Wrong Arithmetic

Imagine an elementary school student who is learning multiplication of fractions and finds that the following appears to be correct: $\frac{1}{4} \times \frac{8}{5} = \frac{18}{45} = \frac{2}{5}$. In other words, the student feels that to do the multiplication, you need merely to combine the digits in the numerator and the denominator to get the right answer since clearly, $\frac{1}{4} \times \frac{8}{5} = \frac{8}{20} = \frac{2}{5}$. Unconvinced with your reasoning that this doesn't work, the student shows you another example, where this method does work, such as $\frac{2}{6} \times \frac{6}{5} = \frac{26}{65} = \frac{2}{5}$. Does this imply that the student has come up with a new method of multiplying fractions? This will certainly give you something to think about. The student may very well say we can flip the two previous fractions and it still works, as we see with $\frac{6}{2} \times \frac{5}{6} = \frac{65}{26} = \frac{5}{2}$, which is a correct result with a wrong procedure. There are 14 such examples, where it works so that you can experience the limits of this weird and *incorrect* multiplication. They are as follows:

$$\frac{1}{4} \times \frac{8}{5} = \frac{18}{45}, \quad \frac{1}{2} \times \frac{5}{4} = \frac{15}{24}, \quad \frac{1}{6} \times \frac{4}{3} = \frac{14}{63}, \quad \frac{1}{6} \times \frac{6}{4} = \frac{16}{64},$$

$$\frac{1}{9} \times \frac{9}{5} = \frac{19}{95}, \quad \frac{4}{9} \times \frac{9}{8} = \frac{49}{98}, \quad \frac{2}{6} \times \frac{6}{5} = \frac{26}{65}$$

Each of these can be flipped to get another seven such examples. Using simple algebra, we can show that these are the *only* seven examples, where the digits are not the same. Let a, b, c, and d be the digits from 1 to 9 so that $\frac{a}{b} \times \frac{c}{d} = \frac{10a+c}{10b+d}$, which then can be reduced to the equation $ac(10b + d) = bd(10a + c)$, which then leads you to the seven examples above and their flips.

Howlers! Reducing Fractions Incorrectly

In our early years of schooling, we learned to reduce fractions. For this, there were specific ways to do it correctly. Some wise guy seems to have come up with a shorter way to reduce some fractions. Is he right? He was asked to reduce a fraction and did it in the following way:

$$\frac{2\not6}{\not65} = \frac{2}{5}$$

That is, he just canceled out the 6s to get the right answer. Is this procedure correct? Can it be extended to other fractions? If so, then we were surely treated unfairly by our elementary school teachers who made us do much more work. Let's look at what was done here and if it can be generalized.

Explanation: In his book, *Fallacies in Mathematics*, E.A. Maxwell refers to the following cancelations as "howlers":

$$\frac{1\not6}{\not64} = \frac{1}{4} \qquad \frac{2\not6}{\not65} = \frac{2}{5}$$

Perhaps when someone did the fraction reductions this way and still got the right answer, it could just make you howl.

Begin by reducing the following fractions to the lowest terms:

$$\frac{16}{64}, \frac{19}{95}, \frac{26}{65}, \frac{49}{98}$$

After you have reduced to the lowest terms each of the fractions in the usual manner, one may ask why it couldn't have been done in the following way:

$$\frac{1\not6}{\not64} = \frac{1}{4}$$

$$\frac{1\not9}{\not95} = \frac{1}{5}$$

$$\frac{2\not6}{\not65} = \frac{2}{5}$$

$$\frac{4\not9}{\not98} = \frac{4}{8} = \frac{1}{2}$$

At this point, you may be somewhat amazed. Your first reaction is probably to ask if this can be done to any fraction composed of two-digit numbers of this sort. Can you find another fraction (comprised of two-digit numbers) where this type of cancelation will work? You might cite $\frac{55}{55} = \frac{5}{5} = 1$ as an illustration of this type of cancelation. This will hold true for all two-digit multiples of 11.

Further explanation: Consider the fraction $\dfrac{10x + a}{10a + y}$.

The above four cancelations were such that when canceling the as, the fraction was equal to $\frac{x}{y}$. Therefore, $\frac{10x+a}{10a+y} = \frac{x}{y}$. This yields $y(10x + a) = x(10a+y)$ and then $10xy+ay = 10ax+xy$, and it follows that $9xy+ay = 10ax$. Thus, $y = \frac{10ax}{9x+a}$.

At this point, we shall inspect this equation. It is necessary that x, y, and a are integers since they were digits in the numerator and denominator of a fraction. It is now our task to find the values of a and x for which y will also be integral. To avoid a lot of algebraic manipulation, you will want to set up a chart which will generate values of y from $y = \frac{10ax}{9x+a}$. Remember that x, y, and a must be single digit integers. In Figure 2.4, we have a portion of the table we constructed. Note that the cases where $x = a$ are excluded since $\frac{x}{a} = 1$.

x\a	1	2	3	4	5	6	...	9
1		$\frac{20}{11}$	$\frac{30}{12}$	$\frac{40}{13}$	$\frac{50}{14}$	$\frac{60}{15}=4$		$\frac{90}{18}=5$
2	$\frac{20}{19}$		$\frac{60}{21}$	$\frac{80}{22}$	$\frac{100}{23}$	$\frac{120}{24}=5$		
3	$\frac{30}{28}$	$\frac{60}{29}$		$\frac{120}{31}$	$\frac{150}{32}$	$\frac{180}{33}$		
4								$\frac{360}{45}=8$
⋮								
9								

Figure 2.4

The portion of the chart pictured above already generated two of the four integral values of y, that is, when $x = 1$, $a = 6$, then $y = 4$, and when $x = 2$, $a = 6$, then $y = 5$. These values yield the fractions $\frac{16}{64}$ and $\frac{26}{65}$, respectively. The remaining two integral values of y will be obtained when $x = 1$ and $a = 9$, yielding $y = 5$, and when $x = 4$ and $a = 9$, yielding $y = 8$. These yield the fractions $\frac{19}{95}$ and $\frac{49}{98}$, respectively. This should convince you that there are only four such fractions composed of two-digit numbers.

You may now wonder if there are fractions composed of numerators and denominators of more than two digits, where this strange type of cancelation holds true. Try this type of cancelation with $\frac{4\cancel{99}}{\cancel{99}8}$. You should find that $\frac{499}{998} = \frac{4}{8} = \frac{1}{2}$.

A pattern is now emerging, and you may realize that:

$$\frac{49}{98} = \frac{499}{998} = \frac{4999}{9998} = \frac{49999}{99998} = \frac{499999}{999998} = \cdots = \frac{4}{8} = \frac{1}{2}$$

$$\frac{16}{64} = \frac{166}{664} = \frac{1666}{6664} = \frac{16666}{66664} = \frac{166666}{666664} = \cdots = \frac{1}{4}$$

$$\frac{19}{95} = \frac{199}{995} = \frac{1999}{9995} = \frac{19999}{99995} = \frac{199999}{999995} = \cdots = \frac{1}{5}$$

$$\frac{26}{65} = \frac{266}{665} = \frac{2666}{6665} = \frac{26666}{66665} = \frac{266666}{666665} = \cdots = \frac{2}{5}$$

Enthusiastic readers may wish to justify these extensions of the original howlers. Readers who, at this point, have a further desire to seek out additional fractions, which permit this strange cancelation should consider the following fractions. They should verify the legitimacy of this strange cancelation and then set out to discover more such fractions.

$$\frac{3\cancel{3}2}{8\cancel{3}0} = \frac{32}{80} = \frac{2}{5}$$

$$\frac{3\cancel{8}5}{8\cancel{8}0} = \frac{35}{80} = \frac{7}{16}$$

$$\frac{1\cancel{3}8}{\cancel{3}45} = \frac{18}{45} = \frac{2}{5}$$

$$\frac{2\cancel{7}5}{7\cancel{7}0} = \frac{25}{70} = \frac{5}{14}$$

$$\frac{1\cancel{6}\cancel{3}}{\cancel{3}2\cancel{6}} = \frac{1}{2}$$

Aside from providing an algebraic application, which can be used to introduce a number of important topics in a motivational way, this topic can also provide some recreational activities. Here are some more of these "howlers":

$$\frac{484}{847} = \frac{4}{7} \qquad \frac{545}{654} = \frac{5}{6} \qquad \frac{424}{742} = \frac{4}{7} \qquad \frac{249}{996} = \frac{24}{96} = \frac{1}{4}$$

$$\frac{48484}{84847} = \frac{4}{7} \qquad \frac{54545}{65454} = \frac{5}{6} \qquad \frac{42424}{74242} = \frac{4}{7}$$

$$\frac{3243}{4324} = \frac{3}{4} \qquad \frac{6486}{8648} = \frac{6}{8} = \frac{3}{4}$$

$$\frac{14714}{71468} = \frac{14}{68} = \frac{7}{34} \qquad \frac{878048}{987804} = \frac{8}{9}$$

$$\frac{1428571}{4285113} = \frac{1}{3} \qquad \frac{2857142}{8571426} = \frac{2}{6} = \frac{1}{3} \qquad \frac{3461538}{4615384} = \frac{3}{4}$$

$$\frac{767123287}{876712328} = \frac{7}{8} \qquad \frac{3243243243}{4324324324} = \frac{3}{4},$$

$$\frac{1025641}{4102564} = \frac{1}{4} \qquad \frac{3243243}{4324324} = \frac{3}{4} \qquad \frac{4571428}{5714285} = \frac{4}{5}$$

$$\frac{4848484}{8484847} = \frac{4}{7} \qquad \frac{5952380}{9523808} = \frac{5}{8} \qquad \frac{4285714}{6428571} = \frac{4}{6} = \frac{2}{3}$$

$$\frac{5454545}{6545454} = \frac{5}{6} \qquad \frac{6923076}{9230768} = \frac{6}{8} = \frac{3}{4} \qquad \frac{4242424}{7424242} = \frac{4}{7}$$

$$\frac{5384615}{7538461} = \frac{5}{7} \qquad \frac{2051282}{8205128} = \frac{2}{8} = \frac{1}{4} \qquad \frac{3116883}{8311688} = \frac{3}{8}$$

$$\frac{6486486}{8648648} = \frac{6}{8} = \frac{3}{4} \qquad \frac{484848484}{848484847} = \frac{4}{7}$$

This problem shows how elementary algebra can be used to investigate a number theory situation, one that is also quite amusing. Mathematics continues to hold some hidden treasures.

Curious Ways to Represent the Number 1,000,000

The number 1,000,000 can be written in many ways as a sum of four *even* numbers. The number 1,000,000 can also be written in many ways as a sum of four *odd* numbers. The question being posed here is as follows: Are there more ways to represent the number 1,000,000 with four even numbers or with four odd numbers (disregarding the order of the summands, which should all be positive)? Or are there equally many ways?

Explanation: The sum of the even numbers is $2a + 2b + 2c + 2d = 1,000,000$, which gives us $a + b + c + d = 500,000$, and implies that one has to count all the possibilities for representing 500,000 as a sum of four natural numbers $a, b, c, d \geq 1$.

The sum of odd numbers can be written as $(2a - 1) + (2b - 1) + (2c - 1) + (2d - 1) = 1,000,000$, which gives us $a + b + c + d = 500,002$, that is, one has to count all the possibilities for representing 500,002 as a sum of four natural numbers $a, b, c, d \geq 1$, and these are more than in the case of even numbers because $500,002 > 500,000$.

We can simplify this by considering the number 10 instead of 1,000,000 where this effect is rather clearer. There is only one possibility regarding 10 as the sum of four even numbers (that is, $2 + 2 + 2 + 4$); however, there are three possibilities for the sum of four odd numbers, namely $1 + 1 + 1 + 7$, $1 + 1 + 3 + 5$, and $1 + 3 + 3 + 3$.

Counting Paths

Counting paths can be a challenge unless a pattern evolves. Consider the grid shown in Figure 2.5 and then determine the number of permitted ways one can go from A to B by moving from square to square either to the right or upwards.

Explanation: Start at A and fill in every square with the number of possibilities to get there. Then, one quickly recognizes that every such number is the sum of the number below and the number to the left. This then allows one to fill the squares with larger numbers without testing the number of ways being indicated. Hence, there are 486 ways to reach square B, as shown in Figure 2.6.

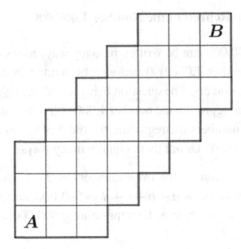

Figure 2.5

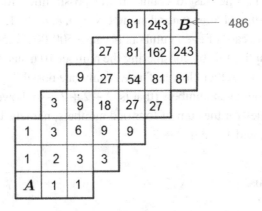

Figure 2.6

A Most Unusual Magic Square

Typically, magic squares are square arrangements of numbers, where the sum of each row, column, and diagonal is the same. However, there are times in most unusual cases where this can be extended from a normal magic square to one that has a surprising result. Consider the 3 × 3 magic square shown in Figure 2.7.

2	7	6
9	5	1
4	3	8

Figure 2.7

In each case, the sums of the rows, columns, and diagonals are 15, which is what is expected in a proper magic square. However, this particular magic square can be modified when its elements are exponents of the number 2 because a new magic square evolves where the products of the rows, columns, and diagonals are the same. We show this in Figures 2.8 and 2.9, where the products of the rows, columns, and diagonals are all 32,768. Thus, we have created an unusual magic square based on a common product emanating from a common sum.

2^2	2^7	2^6
2^9	2^5	2^1
2^4	2^3	2^8

Figure 2.8

4	128	64
512	32	2
16	8	256

Figure 2.9

An ambitious reader may choose to try to create a product magic square using a base other than 2.

A Peculiar Large Number

A huge number n consists of 300 units digits and an unknown number of
zeros. Can this number be a square number? If yes, in which cases, if no,
why not?

Explanation: Recall that a number is divisible by 3, if, and only if, the sum
of its digits is divisible by 3, and a number is divisible by 9, if, and only
if, the sum of its digits is divisible by 9. The sum of the digits of n is 300
since there are 300 units digits, therefore, the number n is divisible by 3. In
order for n to be a square number, the sum of the digits would have to be
divisible by 9 and not by 3 alone, among others. Unfortunately, the sum of
the digits is 300, which is not divisible by 9, and therefore, n itself is not
divisible by 9 and, thus, cannot be a square number.

**The Sum of Consecutive Odd Numbers (Starting at 1) is Always
a Perfect Square**

$$1 + 3 + 5 + \cdots + (2n - 1) = n^2$$

There are numerous proofs for this phenomenon, one of the easiest needs
just a picture (Figure 2.10), not a formal proof by mathematical induction.

Figure 2.10

The following is clear:

(1) Helpful squares appear (starting from a square 1×1 the side length (both width and height) increases in every step by 1, which means that the shape remains a square).

(2) The number of circles increases by 2 from every angle hook to its successor (by 1 on each leg), which means the number of circles — starting from 1 — remains odd.

Hence, this pattern surely continues.

A Remarkable Relationship

The following three equations demonstrate some very beautiful symmetry and equality. Does this pattern continue indefinitely? Or does it end with these three equations?

$$1 + 2 = 3$$
$$4 + 5 + 6 = 7 + 8$$
$$9 + 10 + 11 + 12 = 13 + 14 + 15$$
$$\vdots \qquad \vdots$$

Explanation: If you continue this process, we find that the pattern continues:

$$16 + 17 + 18 + 19 + 20 \quad = 21 + 22 + 23 + 24$$
$$25 + 26 + 27 + 28 + 29 + 30 = 31 + 32 + 33 + 34 + 35$$

Proof: Using $1 + 3 + 5 + \cdots + (2k - 1) = k^2$ (see above), we can be sure that every line, say line number k, begins with k^2, and in this line, $k + 1$ summands are on the left-hand side of the equation and k summands on the right-hand side of the equation (in both cases consecutive natural numbers). So, we have to prove

$$\underbrace{k^2}_{k \cdot k} + \underbrace{(k^2 + 1) + \cdots + (k^2 + k)}_{k \text{ summands}} = \underbrace{(k^2 + k + 1) + \cdots + (k^2 + k + k)}_{k \text{ summands}}$$

But this is easy to prove: Take the k many ks of $k^2 = k \cdot k$ (first summand above) and put in each of the following brackets on the left-hand side one summand k, then we have exactly the right-hand side of the equation.

Another Remarkable Relationship

Here is another surprising relationship to further enhance your appreciation for mathematical wonders:

$$\frac{1}{3} = \frac{1+3}{5+7} = \frac{1+3+5}{7+9+11} = \frac{1+3+5+7}{9+11+13+15} = \cdots = \frac{1}{3}$$

The numerator is the sum of the first k odd numbers, and the denominator is the sum of the next k odd numbers. How might you justify this result without using the formula for the sum of an arithmetic series?

Explanation: Consider the following "proof-without-words" for the case $k = 4$, where

$$\frac{1+3+5+7}{9+11+13+15} = \frac{1}{3}$$

The odd numbers in Figure 2.11, which are $1, 3, 5, 7, 9, 11, 13, 15$, are represented by "right-angle hooks," and there are equally many such angle hooks of light dots and darker dots ($k = 4$ in both cases). And thus, we can see that the light dots are $\frac{1}{3}$ of the darker dots because they are $\frac{1}{4}$ of the whole picture.

Of course, the same works with other values of k! Obviously, such pictures can prove to be significant explanations ("proofs without words") and in many cases, better than using elaborate formulas.

The Paved Paths

Here, we set out to create paved paths consisting of rectangular tiles, shown in Figure 2.12, each of which has a length of 2 ft and a width of 1 ft. The challenges before us are as follows:

- How many possibilities are there to pave a path of length 10 ft and width 2 ft with 10 such rectangular tiles?

- How many possibilities are there to pave a path of length 15 ft and width 2 ft with 15 such rectangular tiles?
- Describe *generally* for the number of possibilities with n such tiles and a path of length n.

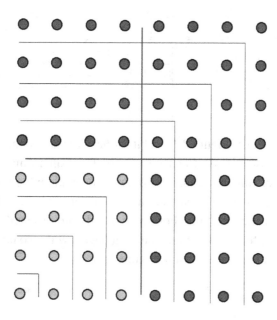

Figure 2.11

Figure 2.12

Explanation: One can easily find the number of possibilities for smaller numbers of n:

n (number of tiles, length of the path)	1	2	3	4	5
$s(n)$ (number of solutions)	1	2	3	5	8

n (number of tiles, length of the path)	Possible paths	Number of paths
1		1
2		2
3		3
4		5
5	We shall leave these for the reader	8
6	We shall leave these for the reader	13

Figure 2.13

If we take the list (Figure 2.13) a bit further, we can see that with $n = 6$, there are 13 possible ways to pave the path. This pattern may remind you of the Fibonacci numbers, and, indeed, this is the underlying principle. The explanation is easy in retrospect but not so easy to find when dealing with the problem for the first time. For every paved path of length n, there are two possibilities for the beginning; either it begins with two horizontal tiles or with a vertical tile, as shown in Figure 2.14.

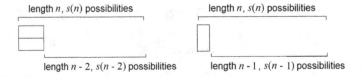

Figure 2.14

In the first case, there are $s(n-2)$ possibilities to pave the remaining path, and in the second case, there are $s(n-1)$ possibilities for the remaining path. Altogether, we can see that the number $s(n)$ can be split up into $s(n-2)$ cases on the one hand and $s(n-1)$ cases on the other hand. Therefore, $s(n) = s(n-1) + s(n-2)$.

Diagonal Properties of a Rectangle

On a graph paper consisting of squares, a rectangle is drawn with its vertices at square intersection points, as shown in Figure 2.15. Consider a rectangle with side lengths m (horizontally) and n (vertically) and then draw one diagonal. The challenge is to determine how many small squares are hit by this diagonal. We then need to establish a general formula. A small square is said to be "hit" by a diagonal if the diagonal passes through its interior; if the diagonal meets only one of its vertices, it is not considered to have been hit by the diagonal.

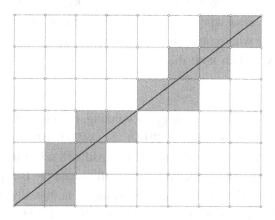

Figure 2.15

Explanation: Let us in the first case assume that m and n are relatively prime. Then (as in Figure 2.15), the diagonal will not pass through vertices of the small squares somewhere in the interior of the rectangle. How many small squares are hit in this case? Every time the diagonal passes one of the *interior $m-1$ vertical* lines, a small square will be hit. Analogously, every time the diagonal passes one of the *interior $n-1$ horizontal* lines, a small square is hit. Together with the first situation, we have $1+(m-1)+(n-1) = m+n-1$ hit small squares.

Next, we shall consider the general case, where m and n are not necessarily relatively prime numbers, with $\gcd(m,n) = d \geq 1$. Then, we divide m and n by d yielding $m' = \frac{m}{d}$ and $n' = \frac{n}{d}$, where m' and n' are again relatively prime. In such a smaller rectangle with dimension $m' \times n'$, we have $m'+n'-1$ hit small squares, as shown in Figure 2.15. Since the rectangle with dimension $m \times n$ has d such "sections" (horizontally or vertically) in which we have smaller rectangles of dimension $m' \times n'$, we have in sum

$d\left(m' + n' - 1\right) = d\left(\frac{m}{d} + \frac{n}{d} - 1\right) = m + n - d$ hit small squares in the general case.

Strategy Game on the Whiteboard to the Year 2017

On a whiteboard are the numbers 1, 2, 3, ..., 2017. David and Lisa alternately delete one number each, until only two numbers remain. David starts. If the sum of the last two numbers is divisible by 8, then David wins, otherwise Lisa wins. Is there a strategy for winning for David or Lisa? If yes, for whom, and how does it work? In which year could this game be played again with the same strategy or perhaps with a similar one?

Explanation: David has a strategy for winning. He thinks of building pairs with a sum 2016: (1, 2015); (2, 2014); ... ; (1007, 1009), that is, if Lisa selects one of these numbers in the pair, David will select the other paired number. Following this procedure, there will remain the pair (1008, 2016) and the number 2017. David's strategy is that if he initially deletes 2017, then all the other pairs remain (with a sum divisible by 8). After each deletion by Lisa, he can then respond by deleting the corresponding 2016-sum-pair number. For example, when Lisa deletes 3, he deletes 2013, and when Lisa deletes 1008, he deletes 2016. Thus, it is guaranteed that after his deletion, the sum of the remaining numbers is always divisible by 8, and he will win!

That works exactly in the same way with all numbers of the form $16k + 1$; the next year that this scheme will work after 2017 is in the year 2033 (16 years later). There, David deletes the last number $16k + 1$ and then thinks of pairs with sum $16k$ such as the following: $(1, 16k - 1)$, $(2, 16k - 2)$, ..., $(8k - 1, 8k + 1)$ and the pair $(8k, 16k)$ — this is the only one not having sum $16k$. And with the analogous strategy, he can guarantee that after his move, the sum of the remaining numbers will be divisible by 8.

With a slight change, it also works for all numbers of the form $8k - 1$, such as in the year 2023. Building pairs with sum 2024 (in general $8k$) yields (1, 2023); (2, 2022); ... ; (1011, 1013) and the number 1012 (in general $4k$), which means that David should initially delete 1012 ($4k$), and then he can answer each of Lisa's moves advantageously (so that the sum of the remaining numbers will always be divisible by 8). For example, if

Lisa deletes 1011, he should delete the corresponding "partner" 1013 since $1011 + 1013 = 2024$.

An Arithmetic Phenomenon

Here is a wonderful opportunity to show the usefulness of algebra, for it will be through algebra that the curiosity will be quenched. Consider the following:

> Select any three-digit number with all digits different from one another. Write all possible two-digit numbers that can be formed from the three digits selected. Then, divide their sum by the sum of the digits in the original three-digit number. Everyone in the audience should get the same answer, 22.

Let's consider the three-digit number 365. Take the sum of all the possible two-digit numbers that can be formed from these three digits: $36 + 35 + 63 + 53 + 65 + 56 = 308$. Then, we get the sum of the digits of the original number, which is $3 + 6 + 5 = 14$. When we are to divide 308 by 14, we get 22, which everyone should have gotten regardless of which original three-digit number was selected.

Explanation: Let's analyze this unusual result, where everyone arrives at the number 22 regardless of which three-digit number was initially selected. We will begin with a general representation of the selected number: $100x + 10y + z$. We now take the sum of all the two-digit numbers taken from the original three digits:

$$(10x + y) + (10y + x) + (10x + z) + (10z + x) + (10y + z) + (10z + y)$$
$$= 10(2x + 2y + 2z) + (2x + 2y + 2z)$$
$$= 11(2x + 2y + 2z)$$
$$= 22(x + y + z)$$

When this value $22(x + y + z)$ is divided by the sum of the digits $(x + y + z)$, the result is 22.

With this algebraic explanation, we ought to get a genuine appreciation as to how nicely algebra allows us to understand arithmetic curiosities.

Understanding a Very Special Six-Digit Number 142857

It has many striking properties:

(a) If the last digit is removed and put as the first digit, the result is five
times the number:

$$142,857 \rightarrow 714,285 = 5 \times 142,857$$

(b) If the first digit is removed and put as the last digit, the result is three
times the number:

$$142,857 \rightarrow 428,571 = 3 \times 142,857$$

(c) If the first two digits are removed and put as the last digits, the result is
two times the number:

$$142,857 \rightarrow 285,714 = 2 \times 142,857$$

(d) If the last two digits are removed and put as the first digits, the result is
four times the number:

$$142,857 \rightarrow 571,428 = 4 \times 142,857$$

(e) If the first (last) three digits are removed and put as the last (first) digits,
the result is six times the number:

$$142,857 \rightarrow 857,142 = 6 \times 142,857$$

Are there other six-digit numbers with the properties shown for the number?
The challenge is to see if there are other such numbers, or, if not, we need
to prove there are no other such numbers.

We will now consider the properties mentioned above and noted as (a)
to (e) to check their veracity:

(a) Let us denote the six-digit number z by $z = 10a + b$, where a is
a five-digit number and b a single digit: $z = \underbrace{a}_{\text{5-digit number}} b =$
$10a + b$. Then, the number $\bar{z}$ with b placed as the first digit is
$\bar{z} = b \underbrace{a}_{\text{5-digit number}} = 100,000b + a$ and the condition $5z = \bar{z}$
can be written as

$50a+5b = 100,000b+a \iff 49a = 99,995b \iff 7a = 14,285b$ (*)
Since 14,285 is not divisible by 7 and we see from (*) that $14,285b$ must be divisible by 7, we can conclude $b = 7$ and $a = 14,285$, altogether $z = 142,857$, which is the only such six-digit number.

(b) Now, the roles of z and $\bar{z}$ are interchanged and the factor 5 is replaced by 3: $3\bar{z} = z$.

$10a + b = 300,000b + 3a \iff 7a = 299,999b \iff a = 42,857b$

Thus, $42,857b$ must have five digits since a has five digits, hence, the only two possible values of b are $b = 1$ and $b = 2$ (with $b \geq 3$, the number $42,857b$ would have more than five digits); we then have two solutions:

$$b = 1: \quad a = 42,857, \quad \bar{z} = 142,857, \quad z = 3\bar{z} = 428,571$$
$$b = 2: \quad a = 85,714, \quad \bar{z} = 285,714, \quad z = 3\bar{z} = 857,142$$

(c) Now, a is a four-digit number and b is the moved two-digit number:

$$z = \underbrace{\quad b \quad}_{\substack{\text{2-digit} \\ \text{number}}} \underbrace{\quad a \quad}_{\substack{\text{4-digit} \\ \text{number}}} = 10,000b + a.$$

Then, the number $\bar{z}$ with b put as the last two digits is

$$\bar{z} = \underbrace{\quad a \quad}_{\substack{\text{4-digit} \\ \text{number}}} \underbrace{\quad b \quad}_{\substack{\text{2-digit} \\ \text{number}}} = 100a+b \text{ and the condition } \bar{z} = 2z \text{ can be written}$$

as $100a+b = 20,000b+2a \iff 98a = 19,999b \iff 14a = 2,857b$ (**)

Since 2,857 is a prime number, it follows from (**) that a is a multiple of 2857 and b is the same multiple of 14. As for corresponding factor k ("the same multiple"), there are only three possibilities $k = \begin{cases} 1 \\ 2 \\ 3 \end{cases}$

because $a = k \cdot 2857$ must stay at four digits: $b = \begin{cases} 14 \\ 28 \\ 42 \end{cases} \quad a = \begin{cases} 2857 \\ 5714 \\ 8571 \end{cases}$

yielding three solutions $z = \begin{cases} 142,857 \\ 285,714 \\ 428,571 \end{cases}$.

(d) In relation to (c), again the roles of z and $\bar{z}$ are interchanged and the factor 4 is replaced by 4: we have $4\bar{z} = z$.

$$400a + 4b = 10{,}000b + a \iff 399a = 9{,}996b \iff 19a = 476b$$

Since 19 is a prime number, it follows that b is a multiple of 19 and a is the same multiple of 476. As corresponding factor k ("the same

multiple"), there are only three possibilities $k = \begin{cases} 3 \\ 4 \\ 5 \end{cases}$ because $b = k \cdot 19$

must have two digits and $a = k \cdot 476$ must have four digits:

$$a = \begin{cases} 1428 \\ 1904 \\ 2380 \end{cases} \quad b = \begin{cases} 57 \\ 76 \quad \text{yielding three solutions} \\ 95 \end{cases}$$

$$\bar{z} = \begin{cases} 142{,}857 \\ 190{,}476 \; , \\ 238{,}095 \end{cases} \quad z = 4\bar{z} = \begin{cases} 571{,}428 \\ 761{,}904 \; . \\ 952{,}380 \end{cases}$$

(e) Now, a and b both are three-digit numbers:

$$z = \underbrace{a}_{\substack{\text{3-digit} \\ \text{number}}} \underbrace{b}_{\substack{\text{3-digit} \\ \text{number}}} = 1000a + b. \text{ Then, the number } \bar{z} \text{ with } a \text{ and } b$$

interchanged is $\bar{z} = \underbrace{b}_{\substack{\text{3-digit} \\ \text{number}}} \underbrace{a}_{\substack{\text{3-digit} \\ \text{number}}} = 1000b + a$ and the condition $6z = \bar{z}$

can be written as $6000a + 6b = 1000b + a \iff 5999a = 994b \iff 857a = 142b$.

Since 857 is a prime number, it follows that $b = 857$ and $a = 142$, thus, the only solution is $z = 142{,}857$.

Remark: The background from number theory (not needed in the formulation of the problem) is the notation of $\frac{i}{7}$ $(i = 1, \ldots, 6)$ as periodical decimal fractions:

$$\frac{1}{7} = 0.\overline{142857}, \; \frac{2}{7} = 0.\overline{285714}, \; \frac{3}{7} = 0.\overline{428571}, \; \frac{4}{7} = 0.\overline{571428},$$

$$\frac{5}{7} = 0.\overline{714285}, \; \frac{6}{7} = 0.\overline{857142}$$

The Ever-Present Number 6174

There are some numbers in our decimal system that have unique characteristics. One such number is 6147. To exhibit this strange characteristic, we
begin by selecting any four-digit number, where the digits are not all the
same. Following the procedure that we provide in the following, using any
such four-digit number, will end up with a number 6174.

(1) Begin by selecting any four-digit number — except one that has all
    digits the same.
(2) Rearrange the digits of the number so that they form the largest number
    possible. (In other words, write the number with the digits in descending
    order.)
(3) Then, rearrange the digits of the number so that they form the smallest
    number possible. (That is, write the number with the digits in ascending
    order. Zeros can take the first few places.)
(4) Subtract these two numbers (obviously, the smaller from the larger).
(5) Take this difference and continue the process, over and over and over,
    until you note something disturbing happening. Don't give up before
    something unusual happens.

You will note that this entertaining exercise will eventually arrive at the
number **6,174** — perhaps after one subtraction or after several subtractions.
Once you arrive at this number 6,174, you will find yourself in an endless
loop, which means that by continuing the process with the number 6,174,
you will continue to end up with 6,174. Remember that you will have begun
with an arbitrarily selected number.

Here is an example of how this works with our arbitrarily selected starting number 3,927:

- The *largest* number formed with these digits is 9,732;
- the *smallest* number formed with these digits is 2,379;
- the difference is 7,353.

Now, using this number, 7,353, we continue the process:

- The largest number formed with these digits is 7,533;
- the smallest number formed with these digits is 3,357;
- the difference is 4,176.

Again, we repeat the process.

- The largest number formed with these digits is 7,641
- The smallest number formed with these digits is 1,467
- The difference is 6,174

When one arrives at 6,174, the number continuously reappears. Note that the largest number that can be formed with these digits is 7,641, the smallest number is 1,467 and the difference as we have seen above is 6,174. Remember, all this began with an *arbitrarily selected* **four-digit** number and will always end up with the number 6,174, which then gets you into an endless loop by continuously getting back to 6,174. This nifty loop was first discovered by an Indian mathematician, Dattathreya Ramachandra Kaprekar (1905–1986), in 1946.[2] We often refer to the number 6,174 as the *Kaprekar constant*.

By the way, just as an aside, the number 6,174 is also divisible by the sum of its digits. That is,

$$\frac{6174}{6+1+7+4} = \frac{6174}{18} = 343$$

Some Variations of the Karprekar Constants

- If you choose a two-digit number (not one with two same digits), then the Kaprekar constant would be 81 and you would end up in a loop of length 5: [81, 63, 27, 45, 09 (, 81)]. There is no loop of length 1 for two-digit numbers.
- If you choose a three-digit number (not one with all same digits), then the Kaprekar constant would be 495 and you would end up in a loop of length 1: [495 (, 495)].

---

[2]Dattatreya Ramchandra Kaprekar announced it at the Madras Mathematical Conference in 1949. He published the result in the paper "Problems involving reversal of digits" in *Scripta Mathematica* in 1953; see also Kaprekar, D. R., "An interesting property of the number 6174." *Scripta Mathematica* 15(1955), 244–245.

- If you choose a four-digit number (not one with all same digits), then the Kaprekar constant would be 6,174 — as we have seen before — and you end up with a loop of length 1: **[6174 (, 6174)]**.
- If you choose a five-digit number (not one of all same digits), then there are three Kaprekar constants: 53,955, 61,974, and 62,964.

   One of length 2:   [53,955, 59,994 (, 53,955)]

   and two of length 4:   [61,974, 82,962, 75,933, 63,954 (, 61,974)]

                          [62,964, 71,973, 83,952, 74,943 (, 62,964)]

You can follow this scheme with six-digit numbers, and you will also find yourself getting into a loop. One number you may find leading you into the loop is 840,852, but do not let this stop you from further investigating this mathematical curiosity.[3] For example, consider the digit sum of each difference. Since the sums of the digits of the subtrahend and the minuend[4] are the same, the difference will have a digit sum that is a multiple of 9. For three- and four-digit numbers, the digit sum is 18. In the case of five- and six-digit numbers, the digit sum appears as 27. It follows that for seven- and eight-digit numbers, the digit sum is 36. Yes, you will find that the digit sum, when this technique is used on nine- and ten-digit number, is 45. You will be pleasantly surprised when you check to see what the digit sum is for even larger numbers.

Another Amazing Curiosity Generated by the Number 1089

This is about a number that has some truly exceptional properties. We begin by showing how the number 1089 just happens to "pop up" when least

---

[3]If you choose a six-digit number (not one with all of the same digits), then there are also three Kaprekar constants: 549,945, 631,764, and 420,876.
Two of length 1: [549,945 (, 549,945)], [631,764 (, 631,764)] and one of length 7: [420,876, 851,742, 750,843, 840,852, 860,832, 862,632, 642,654 (, 420,876)].
If you choose a seven-digit number (not one with all of the same digits), then there is only one Kaprekar constant: 7,509,843.
There is a loop of length 8: [7,509,843, 9,529,641, 8,719,722, 8,649,432, 7,519,743, 8,429,652, 7,619,733, 8,439,552 (, 7,509,843)].

[4]In a subtraction, the number in the *subtrahend* is subtracted from the number in the *minuend* to get the result, referred to as the *difference*.

expected and then we'll take another look at this number. We shall begin by having *you* select any three-digit number, where the units and hundreds digits are not the same and follow the following instructions:

Follow these instructions step by step, while we do it along in the boxes below each instruction.

Choose any three-digit number (where the units and hundreds digits are not the same).

> We will do it with you here by arbitrarily selecting
>
> **825**

Reverse the digits of the number you have selected.

> We will continue here by reversing the digits of 825 to get
>
> **528**

Subtract the two numbers (naturally, the larger minus the smaller).

> Our calculated difference is **825 − 528 = 297**

Once again, reverse the digits of this difference.

> Reversing the digits of 297, we get the number
>
> **792**

Now, add your last two numbers.

> We then add the last two numbers to get 297 + 792 = 1089

Your result should be the same as ours even though your starting number was different from ours.

You will probably be astonished that regardless of which number you selected at the beginning, you got the same result as we did, 1089. How does this happen? Is this a "freak property" of this number? Did we do something devious in our calculations?

Unlike the previous unit, which depended on a peculiarity of the decimal system, this illustration of a mathematical oddity depends on the operations. Before we explore (for the more motivated reader) why this happens, let yourself be impressed with a further property of this lovely number 1089. Let's look at the first nine multiples of 1089:

$$1089 \times 1 = 1089$$

$$1089 \times 2 = 2178$$

$$1089 \times 3 = 3267$$

$$1089 \times 4 = 4356$$

$$1089 \times 5 = 5445$$

$$1089 \times 6 = 6534$$

$$1089 \times 7 = 7623$$

$$1089 \times 8 = 8712$$

$$1089 \times 9 = 9801$$

Do you note a pattern among the products? Look at the first and ninth products (i.e., 1089 and 9801). They are the reverse of one another. The second and the eighth products (i.e., 2178 and 8712) are also reverse of one another. And so, the pattern continues, until the 5th product, 5445, which is a palindromic number (see page 62 for a quick refresher).

Note, in particular, that $1089 \times 9 = 9801$, which is the reversal of the original number. The same property holds for $10989 \times 9 = 98901$, and similarly, $109989 \times 9 = 989901$. You should recognize that we altered the original number, 1089, by inserting a 9 in the middle of the number to get 10989, and extended that by inserting 99 in the middle of the number 1089 to get 109989. It would be nice to conclude from this that each of the following numbers has the same property: 1099989, 10999989, 109999989, 1099999989, 10999999989, and so on.

As a matter of fact, there is only one other number with four or fewer digits where a multiple of itself is equal to its reversal, and that is the number 2178 (which just happens to be 2 × 1089), since 2178 × 4 = 8712. Wouldn't it be nice if we could extend this, as we did with the above example, by inserting 9s into the middle of the number to generate other numbers that have the same property? Yes, it is true that

$$21978 \times 4 = 87912$$
$$219978 \times 4 = 879912$$
$$2199978 \times 4 = 8799912$$
$$21999978 \times 4 = 87999912$$
$$219999978 \times 4 = 879999912$$
$$2199999978 \times 4 = 8799999912$$

and so on.

As if the number 1089 didn't already have enough cute properties, here is another one that (sort of) extends the 1089: We will consider the number 1089 in two parts, the numbers 1 and 89.

Let's see what happens when you take any number and get the sum of the squares of its digits. Then, continue this process of find the sum of the squares of the digits. Each time, curiously enough, you will eventually reach 1 or 89. Take a look at some examples that follow.

We will begin with the number 30. So, we can say that $n = 30$, and we will find the sum of the digits of this number:

$$3^2 + 0^2 = 9, 9^2 = 81, 8^2 + 1^2 = 65, 6^2 + 5^2 = 61, 6^2 + 1^2 = 37,$$
$$3^2 + 7^2 = 58, 5^2 + 8^2 = \mathbf{89}, 8^2 + 9^2 = 145, 1^2 + 4^2 + 5^2 = 42,$$
$$4^2 + 2^2 = 20, 2^2 + 0^2 = 4, 4^2 = 16, 1^2 + 6^2 = 37, 3^2 + 7^2 = 58,$$
$$5^2 + 8^2 = \mathbf{89}, \ldots$$

Once we reached 89, we got into what we call a loop, since we always seem to get back to the number 89, when we repeat the process. Let's try this with the number 31.

So, we will let $n = 31$: $3^2 + 1^2 = 10, 1^2 + 0^2 = \mathbf{1}, 1^2 = \mathbf{1}$

Again, for the number 1, a loop is formed, getting us back to 1 over and over.

We shall now try 32, and we let $n = 32$: $3^2 + 2^2 = 13$, $1^2 + 3^2 = 10$, $1^2 + 0^2 = \mathbf{1}$, $1^2 = \mathbf{1}$

$n = 33$: $3^2 + 3^2 = 18$, $1^2 + 8^2 = 65$, $6^2 + 5^2 = 61$, $6^2 + 1^2 = 37$,

$\quad\quad\quad 3^2 + 7^2 = 58$, $5^2 + 8^2 = \mathbf{89}$, $8^2 + 9^2 = 145$, $1^2 + 4^2 + 5^2 = 42$,

$\quad\quad\quad 4^2 + 2^2 = 20$, $2^2 + 0^2 = 4$, $4^2 = 16$, $1^2 + 6^2 = 37$, $3^2 + 7^2 = 58$,

$\quad\quad\quad 5^2 + 8^2 = \mathbf{89}, \ldots$

$n = 80$: $8^2 + 0^2 = 64$, $6^2 + 4^2 = 52$, $5^2 + 2^2 = 29$, $2^2 + 9^2 = 85$,

$\quad\quad\quad 8^2 + 5^2 = \mathbf{89}$, $8^2 + 9^2 = 145$, $1^2 + 4^2 + 5^2 = 42$, $4^2 + 2^2 = 20$,

$\quad\quad\quad 2^2 + 0^2 = 4$, $4^2 = 16$, $1^2 + 6^2 = 37$, $3^2 + 7^2 = 58$,

$\quad\quad\quad 5^2 + 8^2 = \mathbf{89}, \ldots$

$n = 81$: $8^2 + 1^2 = 65$, $6^2 + 5^2 = 61$, $6^2 + 1^2 = 37$, $3^2 + 7^2 = 58$,

$\quad\quad\quad 5^2 + 8^2 = \mathbf{89}$, $8^2 + 9^2 = 145$, $1^2 + 4^2 + 5^2 = 42$, $4^2 + 2^2 = 20$,

$\quad\quad\quad 2^2 + 0^2 = 4$, $4^2 = 16$, $1^2 + 6^2 = 37$, $3^2 + 7^2 = 58$,

$\quad\quad\quad 5^2 + 8^2 = \mathbf{89}, \ldots$

$n = 82$: $78^2 + 2^2 = 68$, $6^2 + 8^2 = 100$, $1^2 + 0^2 + 0^2 = \mathbf{1}$, $1^2 = \mathbf{1}$

$n = 85$: $8^2 + 5^2 = \mathbf{89}$, $8^2 + 9^2 = 145$, $1^2 + 4^2 + 5^2 = 42$,

$\quad\quad\quad 4^2 + 2^2 = 20$, $2^2 + 0^2 = 4$, $4^2 = 16$, $1^2 + 6^2 = 37$, $3^2 + 7^2 = 58$,

$\quad\quad\quad 5^2 + 8^2 = \mathbf{89}, \ldots$

Now, let's go back to the original oddity of 1089, where we used digit reversals in order to generate 1089 from a selected three-digit number. We assumed that any number we chose would lead us to 1089. How can we be sure? Well, we could try all possible three-digit numbers to see if it works. That would be tedious and not particularly elegant. An investigation of this oddity requires nothing more than some knowledge of elementary algebra. For the reader who might be curious about this phenomenon, we will provide an algebraic explanation as to why it "works."

We shall represent the arbitrarily selected three-digit number, *htu*, as $100h + 10t + u$, where *h* represents the hundreds digit, *t* represents the tens digit, and *u* represents the units digit.

Let $h > u$,[5] which would be the case either in the number you selected or the reverse of it.

In the subtraction, $u - h < 0$; therefore, take 1 from the tens place (of the minuend) making the units place $10 + u$.

Since the tens digits of the two numbers to be subtracted are equal, and 1 was taken from the tens digit of the minuend, then the value of this digit is $10(t - 1)$. The hundreds digit of the minuend is $h - 1$ because 1 was taken away to enable subtraction in the tens place, making the value of the tens digit $10(t - 1) + 100 = 10(t + 9)$.

We can now do the first subtraction

$$
\begin{array}{l}
100(h - 1) + 10(t + 9) + (u + 10) \\
\underline{-(100u + 10t + h)} \\
100(h - u - 1) + 10(9) + u - h + 10
\end{array}
$$

Reversing the digits of this difference gives us

$$100(u - h + 10) + 10(9) + (h - u - 1)$$

Now, adding these last two expressions gives us

$$100(9) + 10(18) + (10 - 1) = \underline{1089}$$

It is important to stress that algebra enables us to inspect the arithmetic process, regardless of the number.

Before we leave the number 1089, we should point out to the reader, who is now so motivated so as to inspect this curious number further, that there is still another oddity, namely, $33^2 = 1089 = 65^2 - 56^2$, which is unique among two-digit numbers.

By this time, you must agree that there is a particular beauty in the number **1089**.

The number 1089 also lends itself to another interesting numerical pattern. When we multiplied 1089 by each of the digits from 1 through 9, the

---

[5]A reminder: The symbol > means "greater than" and the symbol < means "less than".

products have a very curious property: The units digits decreased by 1 each time, beginning with the number 9; the tens digits decreased by 1 each time beginning with the number 8; the hundreds digits increase each time starting with 0; and the thousandths digits increased each time again with the number 1.

Also of note is that the last entry in Figure 2.16 shows that 9801 is a multiple of its reversal 1089. There is only one number of five distinct digits, whose *multiple* is a reversal of the original number. That number is 21,978, since 4 × 21,978 = 87,912, its reverse number. The number 1089 gives rise to similarly structured numbers that provides even more fodder for entertaining an audience.

1 × 1089	1089
2 × 1089	2178
3 × 1089	3267
4 × 1089	4356
5 × 1089	5445
6 × 1089	6534
7 × 1089	7623
8 × 1089	8712
9 × 1089	9801

Figure 2.16

Remember that 1089 × 9 = 9801, which is the reversal of the original number. The same property holds for 10989 × 9 = 98901, and similarly, 109989 × 9 = 989901. You should recognize that we altered the original number, 1089, by inserting a 9 in the middle of the number to get 10989 and extended that by inserting 99 in the middle of the number 1089 to get 109989. It would be nice to conclude from this that each of the following numbers has the same property: 1099989, 10999989, 109999989, 1099999989, 10999999989, and so on.

An Amazing Phenomenon

As we close this chapter, we present a truly amazing numerical relationship. This one will clearly baffle you and impress you with a true wonder about the field of arithmetic. We begin with an equation of numbers equal to 0.

$$123789^2 + 561945^2 + 642864^2 - 242868^2 - 761943^2 - 323787^2 = 0$$

Looking at this relationship, there is nothing particularly strange other than the fact that the numbers are rather large. However, when we delete the 100-thousands place (the left-most digit) from each number, we get the following relationship which remains equal to 0.

$$23789^2 + 61945^2 + 42864^2 - 42868^2 - 61943^2 - 23787^2 = 0$$

When we repeat this process by deleting the left-most digit of each number, we are left with another relationship which is again equal to 0.

$$3789^2 + 1945^2 + 2864^2 - 2868^2 - 1943^2 - 3787^2 = 0$$

When we continue this process of deleting the leftmost digit in each case, the result remains 0.

$$789^2 + 945^2 + 864^2 - 868^2 - 943^2 - 787^2 = 0$$

$$89^2 + 45^2 + 64^2 - 68^2 - 43^2 - 87^2 = 0$$

$$9^2 + 5^2 + 4^2 - 8^2 - 3^2 - 7^2 = 0$$

At this point, you may think that the problem was rigged, as well it might have been. However, we can take the same sequence and repeat the process. But this time deleting the units digit sequentially from each of the numbers, and once again noting that the equation is equal to 0.

$$123789^2 + 561945^2 + 642864^2 - 242868^2 - 761943^2 - 323787^2 = 0$$

$$12378^2 + 56194^2 + 64286^2 - 24286^2 - 76194^2 - 32378^2 = 0$$

$$1237^2 + 5619^2 + 6428^2 - 2428^2 - 7619^2 - 3237^2 = 0$$

$$123^2 + 561^2 + 642^2 - 242^2 - 761^2 - 323^2 = 0$$

$$12^2 + 56^2 + 64^2 - 24^2 - 76^2 - 32^2 = 0$$

$$1^2 + 5^2 + 6^2 - 2^2 - 7^2 - 3^2 = 0$$

If you are not yet impressed enough, then you have an opportunity to consider a bit more wonders by combining the two types of deletions that we have done above, simultaneously! That is, that we will remove the rightmost and leftmost digits at the same time, and once again, amazingly, we achieve a sum of zero with each pair of deletions.

$$123789^2 + 561945^2 + 642864^2 - 242868^2 - 761943^2 - 323787^2 = 0$$

$$2378^2 + 6194^2 + 4286^2 - 4286^2 - 6194^2 - 2378^2 = 0$$

$$37^2 + 19^2 + 28^2 - 28^2 - 19^2 - 37^2 = 0$$

With this wild challenge, we figured to have impressed you sufficiently and can move on beyond arithmetic to the other areas of mathematics, which should also provide a vast variety of entertainment.

Chapter 3

Challenging Problems with Surprising Solutions

To solve problems in everyday life, as in mathematics, there are specific strategies that we tend to use, and we are often not even aware that we are using these specific strategies. Yet, mathematics problems dramatically exhibit some powerful problem-solving techniques. One example of a simple mathematics problem in everyday is a simple decision-making situation; we would say "in the worst-case scenario such and such would be the case," which is, in effect, saying that we are looking for an extreme situation to help solve a problem or concern. The technique of considering extremes is also very enlightening in solving some mathematics problems. Another example is that when traveling from point A to point B, we typically plan a route by determining the roads emanating from the endpoint B and working our path backwards to starting point A. The technique of working backwards to solve problems in mathematics can have some favorable and dramatic effects. Essentially, as we just said, mathematics helps demonstrate the power of specific problem-solving techniques. This chapter presents selected problems that demonstrate how solving problems can be made much simpler by using some of these techniques.

Logical Thinking

Problem: How can you write the numbers 1–100 in a sequence, where no two consecutive numbers have a sum of less than 50?

Solution: Starting with 50 or 49 rather than 51 is also possible. Some logical thinking will lead to the following sequence: 51, 1, 52, 2, 53, 3, 54, 4, 55, 5, . . . , 47, 98, 48, 99, 49, 100, 50.

Logical Thinking

Problem: Show how you can remove only four matchsticks in the square arrangement in Figure 3.1 so that the remaining figure shows only five identically sized squares.

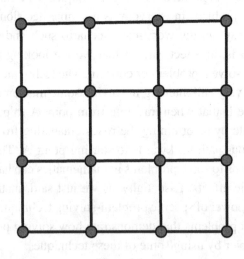

Figure 3.1

Solution: In Figure 3.2, we show the four matchsticks removed leaving exactly five identically sized squares.

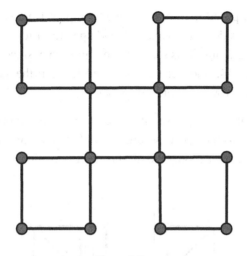

Figure 3.2

Thinking Out of the Box

Problem: Suppose you are presented with a collection of toothpicks arranged, as shown in Figure 3.3, where each of the two outside rows and two outside columns contains 11 toothpicks.

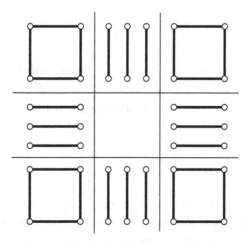

Figure 3.3

The problem posed is to remove one toothpick from each outside row and from each outside column and still end up with 11 toothpicks in each of these rows and columns. This seems to be impossible, since we are actually *removing* toothpicks, and yet we are asked to keep the same number of toothpicks in each row and column, as before.

Solution: In Figure 3.4, we see that we have taken a toothpick from the center portion of each of the rows and each of the columns and then placed another of these center toothpicks in the corner position so that they could be counted twice.

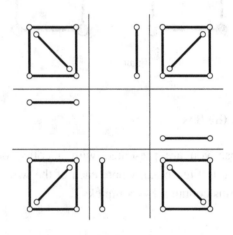

Figure 3.4

Thus, we have achieved our goal of having 11 toothpicks in each of the two rows and each of the two columns. This is a skill that merits attention so that one can go through life by analyzing situations in a more critical fashion.

Thinking Out of the Box

Problem: Consider a regular hexagon formed by 12 matchsticks as shown in Figure 3.5. Show how four matchsticks can be removed and placed elsewhere so that the result will be three equilateral triangles instead of the six equilateral triangles shown in Figure 3.5.

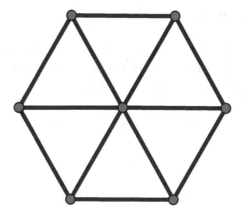

Figure 3.5

Solution: In Figure 3.6, we show with dashed lines the matchsticks that are to be removed and placed accordingly so that the result will be three equilateral triangles.

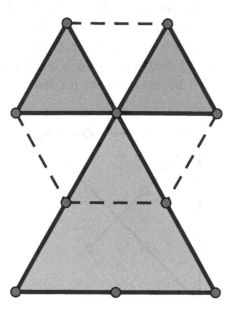

Figure 3.6

Thinking Out of the Box

Problem: Here, you are given a lattice of nine dots as shown in Figure 3.7 and are asked to draw four straight connected lines to touch all the nine dots, without lifting the pencil from the paper.

Figure 3.7

Solution: The typical attempts are to connect the dots along the sides and then note that the four lines have not included the center dot. The trick here is to realize that you have to "think out of the box." In other words, you do not have to be restricted to stay on or within the square formed by the nine dots. In Figure 3.8, we show one possible solution to this trick question.

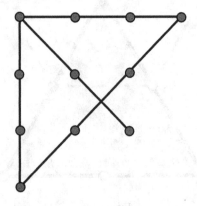

Figure 3.8

Further Thinking Out of the Box

Problem: In Figure 3.9, there are six dots. The challenge is to move one dot to another position so that there will be four rows of dots with three dots in each row.

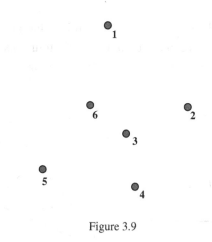

Figure 3.9

Solution: When dot number 6 is moved to the position shown in Figure 3.10, the result is that there are four rows of dots with three dots in each row.

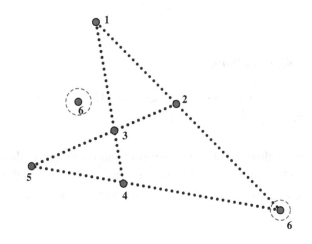

Figure 3.10

Logical and Organized Thinking

Problem: You are given ten coins and asked to place them in such a way that they form five straight lines, each of which contains four coins. There is more than one solution.

Solution: In Figure 3.11, we offer two possible solutions. Note in the diagrams that each of the five lines contains exactly four dots representing the coins. An ambitious reader might seek other solutions.

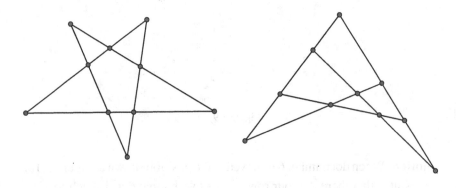

Figure 3.11

Adopting a Different Point of View

Problem: How many numbers in the set {100, 101, 102, 103, ... , 999} do not include the digits 3, 5, 7, and 8?

Solution: Rather than to list all the numbers and then identify those that do not have these four digits included, it would be wise to look at the problem from a different point of view. The numbers that we are considering all have three digits. The first digit cannot include zero and so there are five ways to fill the first digit. There are six ways to fill the second and third digits. Therefore, the number of three-digit numbers that fit our requirement is $5 \times 6 \times 6 = 180$.

Finding a Pattern

Problem: What is the units digit of the number 3^{999}?

Solution: To find the answer using a calculator would probably exceed the capability of most common calculators. Therefore, one clever approach would be to seek a pattern. To do this we will inspect the first 12 powers of 3 as follows:

Values of n	Values of 3^n
1	3
2	9
3	27
4	81
5	243
6	729
7	2,187
8	6,561
9	19,683
10	59,049
11	177,147
12	531,441
13	1,594,323
14	4,782,969
15	14,348,907
16	43,046,721

As we inspect these powers of 3, starting with 3^1, we find that there are repetitions of the units digits every 4 powers ($3 \rightarrow 9 \rightarrow 7 \rightarrow 1 \rightarrow 3 \rightarrow \ldots$). This is clear because the units digit of $3x$ depends only on the units digit of x. So that as we consider the number 3^{999}, we realize that the value of 3^{996} has a units digit of 1, since 996 is divisible by 4, and, therefore, the value of 3^{999}, which is 3 powers further along, must have a units digit of 7.

Logical Reasoning

Problem: On a grid of square spaces, such as that shown in Figure 3.12, an equilateral triangle is to be placed containing dots on its sides so that it has maximum area while at no time containing any of the dots in its interior. How can this equilateral triangle be placed?

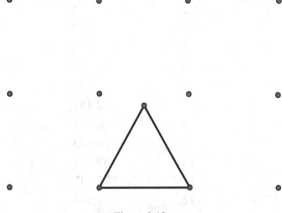

Figure 3.12

Solution: The largest equilateral triangle that can be drawn is one which would include the most dots on its sides, with no dots in its interior, as shown in Figure 3.13. That means that the two sides AB and AC will contain two dots, points D and E, thus, forming equilateral triangle ADE, the sides of which when extended will allow the third side of the triangle to contain points F and G. This produces the largest triangle ABC, which does not have any of the dots in its interior. The logic used here was to maximize the area by maximizing the number of dots on the sides of the triangle.

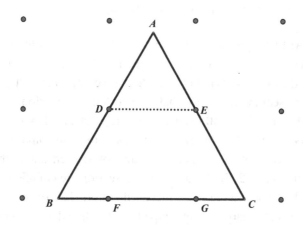

Figure 3.13

Logical Reasoning

Problem: Consider the sequence of numbers 6, 12, 24, 48, 96, 192, 384, . . . , where the difference between consecutive numbers doubles each time. Will there ever be a perfect square among these numbers?

Solution: With logical reasoning, we will realize that a perfect square must have factors in pairs, such as the number 36, whose factors are $2 \times 2 \times 3 \times 3$. Each of the numbers in our given sequence has only one factor of 3 (and can be written as 3×2^n), which will never have a partner number 3, and therefore, there will never be a perfect square in this sequence.

Logical Reasoning

Problem: One of three sources should empty a swimming pool. Which of these three options for emptying the pool is most effective?

Option 1: Three pipes each with a diameter of 20 inches
Option 2: Two pipes each with a diameter of 30 inches
Option 3: One pipe with a diameter of 60 inches

Solution: At first sight, it would seem that there is no difference because the diameter sums in each case are 60 inches. However, logical reasoning tells us that the diameter is not the significant factor here, it is the area of the circular cross-section of the pipe that is important. For Option 1, each of the three pipes has a radius of 10 inches; therefore, each pipe has a cross-section area of 100π for a total of the cross-sections of the three pipes being 300π. For Option 2, each of the pipes has a radius of 15 inches and, therefore, a cross-section area of 225π for a total cross-section area of 450π. However, Option 3, which has a radius of 30 inches, therefore, has a cross-section area of 900π. Hence, the largest single pipe is better than the smaller collection of pipes.

Making a Visual Representation

Problem: Two proofreaders are reading a manuscript in which one proof-reader found 94 errors and the second proofreader found 75 errors. It turns out that 50 errors were found by both proofreaders. How many errors did the manuscript contain?

Solution: There are number of ways to approach this problem, however, a clever technique would be to use a Venn diagram, as we show in Figure 3.14. Since there were 50 errors found by both proofreaders, that would be placed in the intersection of the two circles. One circle, therefore, has 94 errors and the other circle has 75 errors. When we add the three regions, we have $44 + 50 + 25 = 119$ errors.

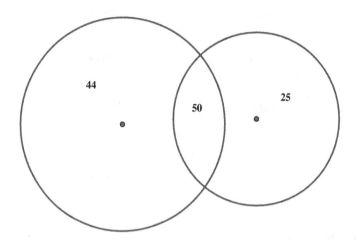

Figure 3.14

Organizing Data

Problem: There are two identical bags shown in Figure 3.15. Bag A has 2 white balls and 3 black balls, and Bag B has 3 white balls and 4 black balls. One bag is chosen at random, and a single ball is removed from it. It was a white ball. What is the probability that the bag chosen was Bag A?

Solution: To organize the data we need to note, as shown in Figure 3.15, that there are 5 balls in Bag A and 7 balls in Bag B. Since the product of the number of balls is $5 \times 7 = 35$, we will imagine having selected 35 balls from each of the two bags. From Bag A, a white ball will be selected 2 out of 5 times or 14 out of 35 tries. From Bag B, a white ball will be chosen 3 out of 7 times, or 15 out of 35 tries. In total, a white ball will be selected $14 + 15 = 29$ times.

Therefore, the probability that it was selected from Bag A is $\frac{14}{29}$.

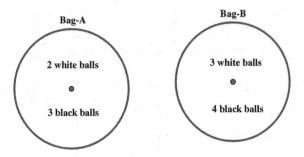

Figure 3.15

Organizing Data

Problem: How many squares are contained in Figure 3.16?

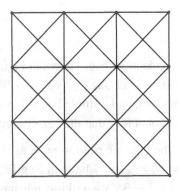

Figure 3.16

Solution: The big mistake would be to just search through the figure and identify as many squares as one can without repeating. However, by organizing data and selecting squares of different sizes and orientations, the process becomes quite simple. We can begin with the obvious complete square and then count the small squares in the interior according to organization and orientation (see Figure 3.17).

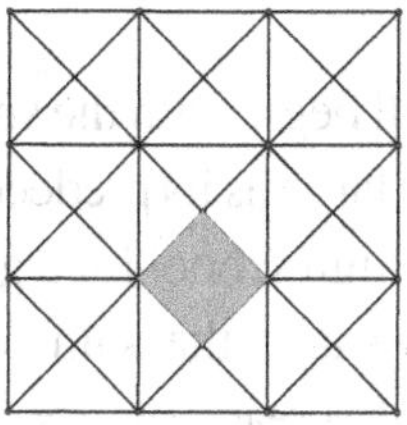

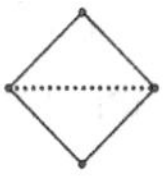

Figure 3.17

Therefore, there are $1 + 9 + 4 + 5 + 12 = 31$ squares in Figure 3.17.

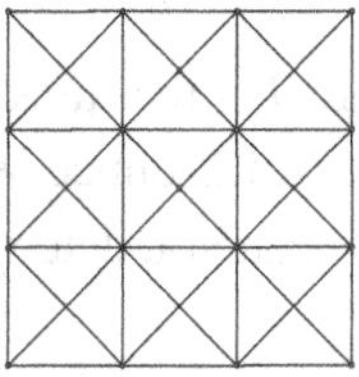

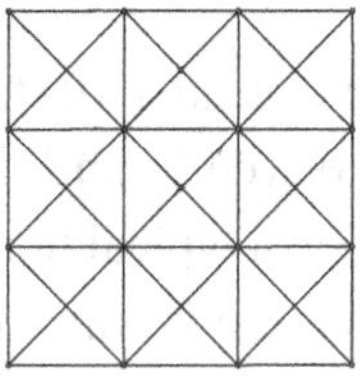

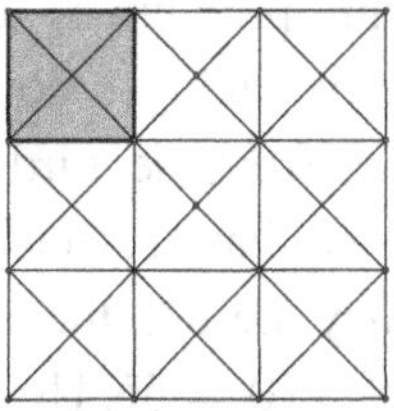

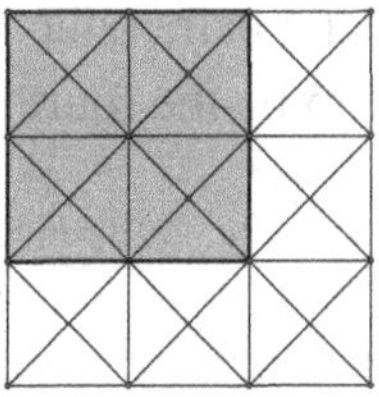

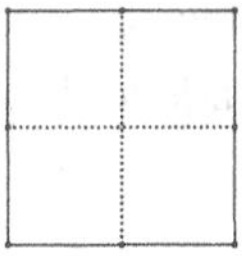

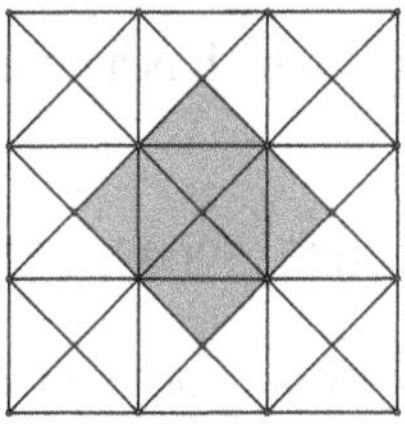

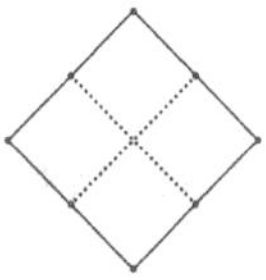

Logical Thinking

Problem: Consider 10 stacks of coins where 9 of the stacks consist of coins of 1-ounce weight and one stack has coins of 2-ounce weight. With one weighing on a typical homestyle scale, how can you determine which stack of coins has the 2-ounce coins?

Solution: The clever way to approach this problem is to find a way to distinguish the stacks. This can be done by taking 1 coin from the first stack, 2 coins from the second stack, 3 coins from the third stack, and so on until you get to the tenth stack where you will have selected 10 coins. If all the coins were the same weight, which they are not, then the total weight would be $1+2+3+4+5+6+7+8+9+10 = 55$. However, when you find that the weight is 59, then it shows that 4 extra ounces were added by one of the stacks, which, in this case, had to be the fourth stack since that would have added 4 extra ounces.

Pattern Recognition

Problem: Consider the following sequence, which is a geometric sequence of multiples of 3:

$$1, 3, 9, 27, 81, 243, 729, 2187, 6561, 19683, 59049, 177147, 531441$$

The challenge here is to determine how to get the sum of the sequence without the normal addition process.

Solution: If we search for a pattern, we will begin with the smallest numbers and note that each number is three times as large as its predecessor. However, in the absence of the formula presented in high school for a geometric series, we can search for a pattern. Suppose we take the sum of the first three numbers $1+3+9 = 13$, which we can also reach by $9 \times 3 - 1 = 26$, and then $26 \div 2 = 13$. To see if this pattern works, we can try it by taking the sum of the first four numbers, $1+3+9+27 = 40$, which could also be reached by multiplying $27 \times 3 - 1 = 80$, then $80 \div 2 = 40$. Now that we see a pattern, we can apply this procedure, which appears to determine the sum of a geometric series by applying it to the original challenge. Therefore, we begin by taking $531,441 \times 3 - 1 = 1,594,322$, which we then divide by 2 to get 797,161, which is the sum we originally sought. We see that discovering a pattern can be a very useful technique, although this approach is not a proof that the recognized pattern is valid in the general case.

Pattern Recognition

Problem: The challenge here is to determine the sum of the digits of all the numbers from 1 to 1,000,000.

Solution: One way to answer this question is to begin by summing the digits of consecutive numbers starting with 1 (see Figure 3.18).

The number	The Digit Sum
1	1
2	2
3	3
. . .	. . .
35	$3 + 5 = 8$
36	$3 + 6 = 9$
37	$3 + 7 = 10$
38	$3 + 8 = 11$
etc.	

Figure 3.18

This does not seem to be a very efficient, or elegant, way to proceed. To determine this sum, rather than simply trying to add the digits of all the numbers, we will use a better arrangement of the numbers and see if we can obtain the sum without actually doing all this tedious addition.

Let's consider the list of the numbers from 0 to 999,999 in two directions: ascending order and descending order (see Figure 3.19).

We will leave the number 1,000,000 till a bit later. Since every pair of numbers has a digit sum of 54, then the million pairs have a digit sum of $54 \times 1,000,000$. The two columns have the same numbers, so the sum of the digits in one column is $27 \times 1,000,000$. We must now add the digit sum of the last number, 1,000,000, which is 1.

Ascending order	Descending order	Digit Sum of Each Pair of Numbers
0	999 999	54
1	999 998	54
2	999 997	54
3	999 996	54
...	...	...
127	999 872	54
...	...	...
257 894	742 105	54
...	...	...
999 997	2	54
999 998	1	54
999 999	0	54

Figure 3.19

Therefore, the digit sum of all numbers from 1 to 1,000,000 is $1,000,000 \times 27 + 1 = 27,000,001$.

From this example, you can see that arithmetic is more than just doing the basic operations of addition, subtraction, multiplication, and division — it requires a bit of thinking as well.

Alternate solution

Let us determine how often the digit 1 is written in the numbers from 0 to 999,999. In every block of ten numbers, namely, $0-9, 10-19, \ldots$ $999,990-999,999$, the digit 1 is written once as a units digit. Since there are 100,000 such blocks of ten numbers, the digit 1 appears 100,000 times as units digit. It appears as a tens-digit also 10 times in every block of hundreds, such as $0-99, 100-199, \ldots, 999,900-999,999$, which yields the digit 1 another 100,000 times. Furthermore, the digit 1 appears another 100 times as a hundreds digit in every block of thousands. This pattern continues for each other power of 10 block, until we reach 100,000. Therefore, since we have 6 such groups of powers of 10, we have the digit 1 appearing 600,000 times. The same holds true for the other digits 2 through 9. So the digit sum is $(1 + 2 + \cdots + 9) \times 600,000 = 27,000,000$. Finally, the last digit 1 from 1,000,000 yields 27,000,001.

Adopting a Different Point of View

Problem: Of the many strategies that are available to us to solve mathematical problems, the one that allows us to avoid "running into the wall," namely, avoiding frustration, is that of approaching the problem from a different point of view. We present here a problem that clearly illustrates how the most common method leads to a correct answer, but it is cumbersome, and more prone to a possible arithmetics error. However, there is a dramatically different, but far simpler, solution method. The problem is as follows: At a school with 25 classes, each of these classes has a basketball team to compete in a schoolwide tournament. In this tournament, a team that loses one game is immediately eliminated. The school only has one gymnasium, and the principal of the school would like to know how many games will be played in this gymnasium in order to get a winner.

The typical solution to this problem could be to simulate the actual tournament by beginning with 12 randomly selected teams playing against a second group of 12 teams, with one team drawing a bye — that is, passing up a game. This would then continue with the winning teams playing against each other as shown here:

Any **12 teams** vs. any other **12 teams,** which leaves **12 winning teams** in the tournament.

6 winners vs. **6 other winners,** which leaves **6 winning teams** in the tournament.

3 winners vs. **3 other winners,** which leaves **3 winning teams** in the tournament.

3 winners + 1 team (which drew a bye) **= 4 teams.**

2 remaining **teams** vs. **2** remaining **teams,** which leaves **2 winning teams** in the tournament.

1 team vs. **1 team** to get a **champion!**

Now counting (summing up) the number of games that have been played (Figure 3.20), we get that the total number of games played is $12 + 6 + 3 + 2 + 1 = 24$.

Teams playing	Games played	Winners
24	12	12
12	6	6
6	3	3
3+ 1 bye = 4	2	2
2	1	1

Figure 3.20

This seems like a perfectly reasonable method of solution and is certainly a correct one. Approaching this problem from a different point of view would be vastly easier by considering the losers, rather than winners, which is what we did in the previous solution. In that case, we ask ourselves, how many losers must there have been in this competition in order to get one champion? Clearly, there had to be 24 losers. To get 24 losers, there needed to be 24 games played. And with that the problem is solved. Looking at the problem from an alternative point of view is a curious and often clever approach that can be useful in a variety of contexts.

Another alternative point of view would be to consider these 25 teams with one of them — only for our purposes — being considered a professional basketball team that would be guaranteed to win the tournament. Each of the remaining 24 teams would be playing the professional team only to lose. Once again, we see that 24 games are required to get a champion. This should demonstrate the power of this problem-solving technique of considering the problem from another point of view.

Solving a Simpler Analogous Problem

Problem: Our challenge here is to determine the relationship between the lengths of the four segments, $a, b, c,$ and d, joining the randomly selected point E in rectangle $ABCD$, which is shown in Figure 3.21.

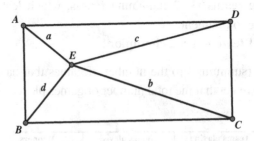

Figure 3.21

Solution: At first view, one might be a bit confused as how to approach this problem. If the point E were in the center of the rectangle, then the problem would be rather simple since the four line segments would be of equal length. Furthermore, if the point E were at the center of the rectangle, the Pythagorean theorem would also be helpful. However, we can invoke the Pythagorean theorem by drawing a horizontal and vertical line through E to form four rectangles, as shown in Figure 3.22. Each of these rectangles has a diagonal which is the hypotenuse of right triangles.

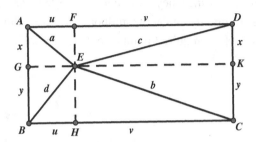

Figure 3.22

Therefore, we will apply the Pythagorean theorem for each of our four initial segments: $a^2 = x^2 + u^2, b^2 = y^2 + v^2, c^2 = x^2 + v^2$, and $d^2 = y^2 + u^2$. A quick overview tells us that since $a^2 + b^2 = (x^2 + u^2) + (y^2 + v^2)$ and $c^2 + d^2 = (x^2 + v^2) + (y^2 + u^2)$, we can clearly see that $a^2 + b^2 = c^2 + d^2$, which shows how adopting a different point of view leads us to a successful response.

Solving a Similar Analogous Problem

Problem: Here we are asked to find the center of gravity of the shape shown in Figure 3.23. In other words, we need to find the point in this shape consisting of rectangles where it could balance on a pin.

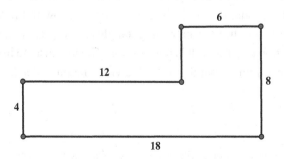

Figure 3.23

Solution: The balancing point of various geometric figures should be known. Perhaps it is intuitive that a rectangle would be balanced at the point of intersection of the diagonals. This figure is comprised of two equal-area rectangles as we can see in Figure 3.24. We draw the diagonals of each of the two rectangles, and since they are equal in area (48 square units), we join their two centers of gravity and at the midpoint, M, of the line segment AB, and we have located the required point of balance for the original figure.

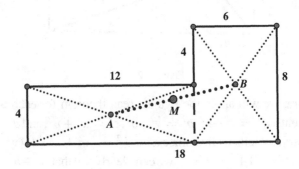

Figure 3.24

An extension: This technique (just taking the midpoint of A and B) would not be possible if the rectangles in question were not of equal area. One would have to find a point dividing the line segment AB in the ratio ba, if the areas of the left and right rectangles were a and b. But in such a case, there is still an easier solution: We locate the centroids[1] of two pairs of rectangles and note that the centroid G must lie on both line segments connecting the partial centroids. Therefore, as we show in Figure 3.25, the pairs of centroids of the two pairs of rectangles are G_1, G_2 and G_3, G_4 and their intersection point is G. It is possible that G could also lie in the exterior of the L-shaped region, but this would obviate the pin balancing.

---

[1]The centroid is the center of gravity or balancing point of a plane figure.

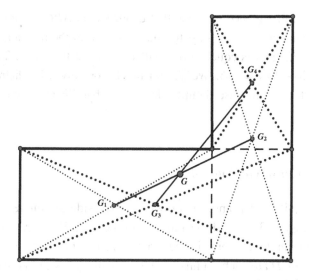

Figure 3.25

Solving a Simpler Analogous Problem

Problem: In Figure 3.26, the ladder is shown with equally spaced steps. The top step has length 20 and the bottom step has length 90. We are asked to find the length of the step marked *EF*.

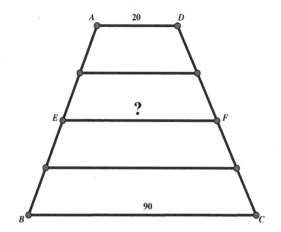

Figure 3.26

Solution: At first glance, we note that quadrilateral *ABCD* is a trapezoid. A simpler analogous problem is to find the length of the median, which in this case is *EF* and is equal to one-half the sum of the bases. Therefore, $EF = \frac{1}{2}(20 + 90) = 55$. We should not let ourselves be distracted by the other steps, rather use them to determine that *EF* is a median of the trapezoid.

Adopting a Different Point of View

When an equilateral triangle is drawn on each side of square *ABCD*, as shown in Figure 3.27, and the remote vertices, *E*, *F*, *G*, and *H*, of the equilateral triangles are joined, a quadrilateral *EFGH* is formed. If the sides of square *ABCD* are each 4 units long, our challenge is to find the area of quadrilateral *EFGH*.

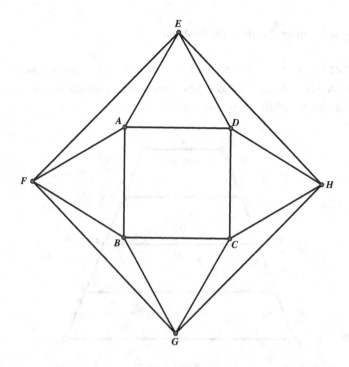

Figure 3.27

Solution: Typically, one is tempted to find the sum of the areas of the equilateral triangles and of the square and the four congruent isosceles triangles, *AEF*, *BFG*, *CGH*, and *HDE*. However, adopting a different point of view would have us consider drawing *FH*, intersecting *AB* and *CD* at points *K* and *L*, respectively. Draw *EM*, where *M* is the midpoint of *KL*, and *EM* intersects *AD* at point *N*. Using the Pythagorean theorem (or recognizing that there are 30°, 60°, and 90° triangles), we get segments of lengths noted in Figure 3.28. It is now rather trivial to find the area of the quadrilateral *EFGH* by finding the area of triangle *EFH*, which is $\frac{1}{2}\left(2\sqrt{3}+4+2\sqrt{3}\right)\left(2+2\sqrt{3}\right) = 8\sqrt{3}+16$. Thus, the area of the quadrilateral *EFGH* is double the area of triangle *EFH* or $16\sqrt{3}+32$.

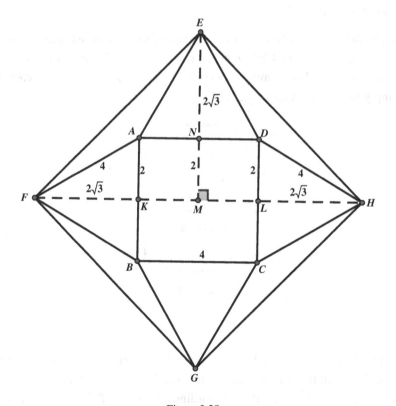

Figure 3.28

Working Backwards

Problem: Charlie has an 11-liter can and a 5-liter can. Charlie's problem is to determine how he can measure exactly 7 liters of water using these cans.

A Common Solution

Most people will simply guess at the answer and keep "pouring" back and forth in an attempt to arrive at the correct answer, a sort of "unintelligent" guessing and testing.

A cleverer solution

The problem can be solved in a more organized manner by using the strategy of working backwards. Charlie needs to end up with 7 liters in the 11-liter can, leaving a total of 4 empty liters in the can. But how can Charlie capture 4 empty liters? (see Figure 3.29).

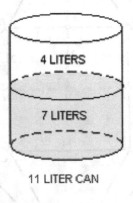

4 LITERS

7 LITERS

11 LITER CAN

Figure 3.29

To obtain 4 liters, he must leave 1 liter in the 5-liter can. Now, how can he obtain 1 liter in the 5-liter can? He fills the 11-liter can and pours from it twice into the 5-liter can, then discarding the water. This leaves 1 liter in the 11-liter can (see Figure 3.30).

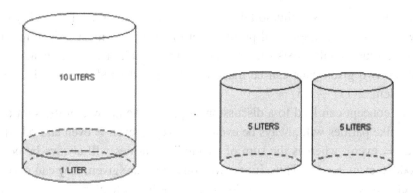

Figure 3.30

He then pours the 1 liter into the 5-liter can. Then he fills the 11-liter can and pours off the 4 liters needed to fill the 5-liter can. This leaves the required 7 liters in the 11-liter can, which is what he wanted to originally have (see Figure 3.31).

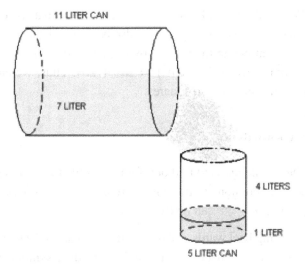

Figure 3.31

Note that problems of this sort do not always have a solution. That is, if you wish to construct additional problems of this sort, you must bear in mind that a solution only exists when the difference of multiples of the capacities of the two given cans can be made equal to the desired quantity. In this problem, $2(11) - 3(5) = 7$.

This concept can lead to a discussion of parity. We know that the sum of two like parities will always be even (i.e., even + even = even and odd + odd = even), whereas the sum of two unlike parities will always be odd (odd + even = odd). Thus, if two even quantities are given, they can never yield an odd quantity. Further discussion can be particularly fruitful as it provides much-needed insight into some valuable number properties and concepts.

Working Backwards

Problem: Here we have a seemingly simple challenge: If $x = a^2 + b^2$, can $2x$ be the sum of two squares as well?

Solution: We would like to know if $2x = 2a^2 + 2b^2$ can also be true from the given value x. In other words, can $2x$ also be a sum of squares? Working backwards, we can create the sum of two squares $(a - b)^2 + (a + b)^2 = (a^2 - 2ab + b^2) + (a^2 + 2ab + b^2) = 2a^2 + 2b^2$. Thus, we have shown how $2x$ can be the sum of two squares.

Working Backwards

Problem: Our challenge is to find out if it is possible to find integer values for a, b, and c, where none of these is a zero or a perfect square so that the equation $\sqrt{a} + \sqrt{b} = \sqrt{c}$ can be true.

Solution: Working backwards, we realize that we can find a solution if the radicals can be converted to be the same under the radical symbol: a "kernel." Therefore, we must find values of a, b, and c that will generate "equal radical kernals." One such is 8, 18, and 50, so that $\sqrt{8} + \sqrt{18} = \sqrt{50}$. All these radicals can be expressed as multiples of $\sqrt{2}$ so that $\sqrt{8} + \sqrt{18} = \sqrt{50}$ can be written as $2\sqrt{2} + 3\sqrt{2} = 5\sqrt{2}$. With this technique, we can find infinitely many such solutions. Let n be not a perfect square (as the common "kernel" of all the radicals), then take $a = a_1^2 n$, $b = b_1^2 n$, and then consequently $c = (a_1 + b_1)^2 n$ with arbitrary natural numbers a_1, b_1. This surely yields a solution since then we have $\sqrt{a} + \sqrt{b} = a_1\sqrt{n} + b_1\sqrt{n} = (a_1 + b_1)\sqrt{n} = \sqrt{c}$.

Working Backwards

Problem: In Figure 3.32, we are asked why the sum of the angles is as follows:

$$\angle A + \angle B + \angle C + \angle D = \angle AED + \angle BFC$$

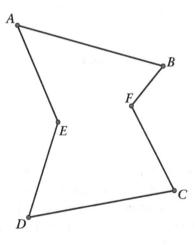

Figure 3.32

Solution: By working backwards, we note that we would need to get a polygon with which we are familiar. That could be a quadrilateral. If we draw line EF, we will have created two quadrilaterals, namely, $ABFE$ and $CDEF$. To make matters simpler, we have noted in Figure 3.33 the four angles at points E and F. In quadrilateral $ABFE$, the sum of the angles $\angle A + \angle B + \angle n + \angle m = 360°$. Yet, along line EF, we get $\angle x + \angle m + \angle y + \angle n = 360°$

or $\angle m + \angle n = 360° - (\angle x + \angle y)$. Therefore, by substitution $\angle A + \angle B + = 360° - (\angle n + \angle m) = 360° - (360 - (\angle x + \angle y)) = \angle x + \angle y$.

In a similar fashion, working backwards with quadrilateral $CDEF$, we get $\angle C + \angle D = \angle w + \angle z$. Thus, by addition we have $\angle A + \angle B + \angle C + \angle D = \angle x + \angle y + \angle w + \angle z = \angle AED + \angle BFC$.

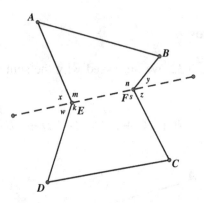

Figure 3.33

Working Backwards

Problem: Here is a very simple-sounding problem, which typically could be solved with a good bit of algebra. However, you will be entertained by a classic example of how working backwards can make the solution trivial. You are given that the sum of two numbers is 2, and the product of these same two numbers is 5. The problem is to find the sum of the reciprocals of these two numbers.

A common approach

The problem immediately suggests forming two equations in two variables:

$$x + y = 2$$

$$xy = 5$$

These two equations can be solved simultaneously by using the quadratic formula, which is $x = \frac{-b \pm \sqrt{b^2 - 4ac}}{2a}$, for $ax^2 + bx + c = 0$. However, the method yields complex values for both x and y, namely, $1 + 2i$ and $1 - 2i$. Following the requirements of the original problem, we now need to take the sum of the reciprocals of these two roots.

$$\frac{1}{1 + 2i} + \frac{1}{1 - 2i} = \frac{(1 - 2i) + (1 + 2i)}{(1 + 2i)(1 - 2i)} = \frac{2}{5}$$

We should emphasize here that there is nothing wrong with this method, it is just not the most elegant way to solve this problem.

A cleverer solution

Before embarking on a problem, it usually makes sense to step back from it and see what is being required. Curiously, this problem is not asking for the values of x and y but rather the sum of the reciprocals of these two numbers. That is, we seek to find $\frac{1}{x} + \frac{1}{y}$. Using a strategy of working backwards, we could ask ourselves from what might this sum have come. Adding these two fractions could give us the answer. Therefore, $\frac{1}{x} + \frac{1}{y} = \frac{x+y}{xy}$. At this point, the required answer is immediately available to us since we know the sum of the numbers is 2, and the product of the numbers is 5, we merely substitute these values in the last fraction to get $\frac{1}{x} + \frac{1}{y} = \frac{x+y}{xy} = \frac{2}{5}$, and our problem is solved.

Working Backwards

Problem: Another challenge of similar nature is to find the sum of the squares of two numbers, where the sum of the two numbers is 10 and the product of these same two numbers is 20.

Solution: As with the previous challenge, the common way to approach this problem is to find the two numbers and then find the sum of the squares. However, with the numbers offered here, this would be a rather tedious task. Yet, this problem can be approached by working backwards. Since we need not determine the two numbers but rather the sum of their squares, we can begin by referring to the two numbers as a and b. Working backwards, we can find the sum of the squares of these two numbers, namely, $a^2 + b^2$.

By taking the sum of the two numbers and squaring, we get $(a+b)^2 = a^2 + b^2 + 2ab$. This can be written as $(a+b)^2 - 2ab = a^2 + b^2$. Thus, by working backwards from our desired conclusion, we never needed to find the actual numbers as we have the sum of squares in terms of the sum of the two numbers and the product of the two numbers, that is, $a^2 + b^2 = (10)^2 - 2(20) = 100 - 40 = 60$. Working backwards has been a great timesaver!

Adopting a Different Point of View

Problem: Here is a challenge which directs us to approach the problem from an unusual point of view. We are given that $x^2 + y^2 = 1$ and $x + y = 1$ and we are asked to find the value of $x^7 + y^7$.

Solution: Typically, an approach would be to try to multiply these two equations and hopefully come up with something that is workable. However, using a different point of view, we can inspect the value of $x + y$. For starters, $\underbrace{x^2 + y^2}_{1} = \underbrace{x^2 + 2xy + y^2}_{=(x+y)^2=1}$, therefore, $2xy = 0$. This implies that either x or y must equal 0. Thus, either x^2 or y^2 must equal 1. Consequently, $x^7 + y^7 = 1$.

Considering Extreme Cases

Problem: When working with infinity unusual aspects arise, which we will appreciate in the challenge to find the positive value of x that satisfies the equation: $x^{x^{x^{x^{\cdot^{\cdot^{\cdot}}}}}} = 2$.

Solution: At first glance, most people would be overwhelmed by the concept of infinity and not know how to approach the problem. We could look at this as being somewhat of an extreme situation by noting that there is an infinite number of x's in this series or tower of powers. Eliminating one of the x's would not have any effect on the end result because of the nature of infinity. Therefore, by removing the first x, we find that all those remaining in the tower of x's must also equal 2. This then permits us to rewrite this equation as $x^2 = 2$. It then follows that $x = \pm\sqrt{2}$. If we remain in the set of positive real numbers, then the answer is $x = \sqrt{2}$.

In the following, you can see how the successive increases get ever closer to 2:

$$\sqrt{2} = 1.414213562\ldots$$

$$\sqrt{2}^{\sqrt{2}} = 1.632526919\ldots$$

$$\sqrt{2}^{\sqrt{2}^{\sqrt{2}}} = 1.760839555\ldots$$

$$\sqrt{2}^{\sqrt{2}^{\sqrt{2}^{\sqrt{2}}}} = 1.840910869\ldots$$

$$\sqrt{2}^{\sqrt{2}^{\sqrt{2}^{\sqrt{2}^{\sqrt{2}}}}} = 1.892712696\ldots$$

$$\sqrt{2}^{\sqrt{2}^{\sqrt{2}^{\sqrt{2}^{\sqrt{2}^{\sqrt{2}}}}}} = 1.926999701\ldots$$

$$\ldots$$

And so, we have a surprisingly simple solution to a very complicated-looking problem. Making sure of the existence of a solution is often crucial. Here, we were lucky because the assumed solution exists. But if we wanted to solve with the "same technique" $y^{y^{y^{y^{\cdots}}}} = 4$, we would get $y^4 = 4 \Rightarrow$ $y = \sqrt[4]{4} = \sqrt{2}$; but we know from above that $\sqrt{2}^{\sqrt{2}^{\sqrt{2}^{\cdots}}} = 2$ (and not 4). The mistake here is to assume that there is a solution and write $y^4 = 4$ instead of the original equation. We must be careful in posing problems that cannot be generalized.

Considering Extreme Cases

Problem: Maximizing the area of the triangle by drawing a line through given lines can be an interesting challenge. In Figure 3.34, point P is situated between lines AB and AC. When a line is drawn through P intersecting AB and AC, a triangle AEF is created. How might this line be drawn through point P in order to yield the smallest area for triangle AEF?

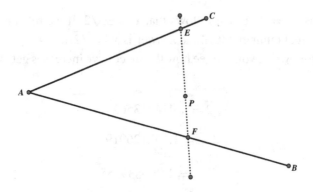

Figure 3.34

Solution: Considering extreme cases, the line *EF* could be placed in such a way that it intersects *AC* at a very distant position. In Figure 3.34, the point *E* could be as far as desired as long as it intersects *AB* and point *P*. This could lead to an extremely large triangle. Therefore, in order to avoid this extreme situation, we can conjecture that the smallest triangle could be formed when point *P* is the midpoint of the line segment *XY*, as we show in Figure 3.35. We claim that under these circumstances, triangle *AXY* has the smallest area. There is an easy way to prove this. Let *RS* be another line through *P* and *XQ* ∥ *RY*, then △*XPQ* ≅ △*RYP* and the area of triangle *ARS* using the line *PR* is larger than the area of triangle *AXY* (the difference being the area of the triangle *XQS*). Therefore, using the line *XY* produces the smallest triangle area.

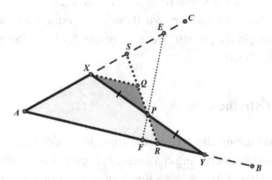

Figure 3.35

Considering Extreme Cases

Problem: Now having been challenged to find the smallest possible triangle under given circumstances, our next challenge is to maximize the area of triangle ABC as we are given three lines AP, BP, and CP, as shown in Figure 3.36, with lengths $AP = 3$, $BP = 5$, and $CP = 7$. How should these three line segments be placed so that a triangle ABC has maximum area?

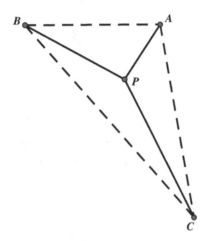

Figure 3.36

Solution: To maximize the area, we need to consider various extreme cases. Suppose we begin by leaving points A, B, and P in their current position for the moment and look to maximize the area of triangle ABC by moving point C to a maximum position. That would be when CP would be perpendicular to AB, as shown in Figure 3.37. Any other position for point C would produce a shorter altitude to line AB and, thereby, a smaller area.

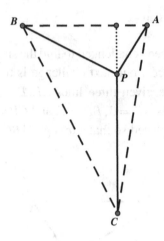

Figure 3.37

Repeating this procedure, by extending each of *BP* and *AP* to be perpendicular to *AC* and *BC*, respectively, as shown in Figure 3.38, will produce the maximum area triangle, as we maximized the length of each of the altitudes.

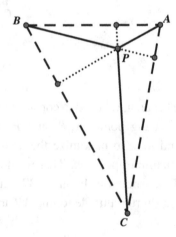

Figure 3.38

Considering Extreme Cases

Problem: In Figure 3.39, we show a square of side length 9 and an isosceles right triangle whose equal sides have length 12 and is so situated that the vertex of the right angle is at the center of the square. The problem here is to find the area of the shaded quadrilateral region.

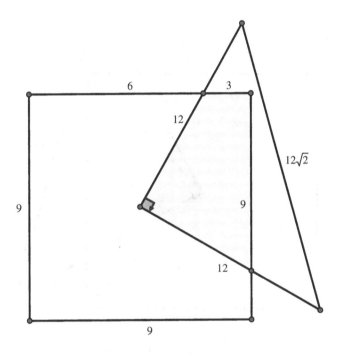

Figure 3.39

Solution: The first attempt might be to try to draw various line segments and then try to find the areas of the square and of the isosceles right triangle. This is a mere distraction. The technique here is to take an extreme situation by rotating the triangle about the right-angle vertex until the sides of the triangle contain the two vertices of the square, as shown in Figure 3.40. It can be easily shown that the two shaded triangles are congruent, and so the area of the original quadrilateral is, therefore, equal to the area in the square formed by the two equal sides of the isosceles right triangle and the side of the square. That area is exactly $\frac{1}{4}$ of the area of the square. Therefore, the area of that original quadrilateral is then $\frac{1}{4}$ of the area of the square which is $\frac{1}{4} \cdot 81 = 20\frac{1}{4}$.

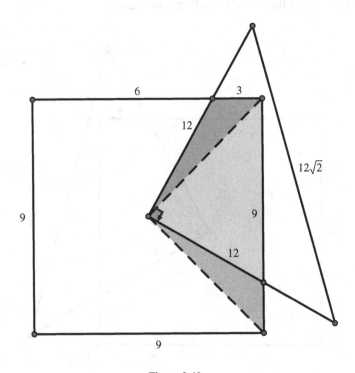

Figure 3.40

Adopting a Different Point of View

Problem: Beginning with a randomly drawn triangle ABC, a square is constructed on each of its sides. Referring to Figure 3.41, we need to compare the area of each of the three shaded triangles, $\triangle ADJ, \triangle BGH$, and $\triangle CEF$, to the area of $\triangle ABC$.

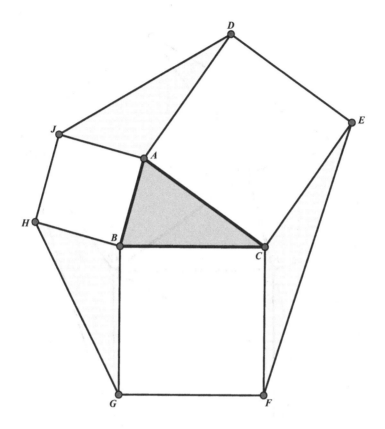

Figure 3.41

Solution: Using one of the area formulas to find the areas of each of these four triangles would be quite challenging, so we will adopt a different point of view. Suppose we focus on comparing the areas $\triangle ABC$ and $\triangle BHG$. Since there are two right angles at point B, we know that $\angle HBG$ and $\angle ABC$ are supplementary. Therefore, if we rotate $\triangle BHG$ 90° about point B so that BG' coincides with BC as we show in Figure 3.42 solution, we find that ABH' is a straight line, where $AB = BH'$. Thus, area $\triangle ABC$ = area $\triangle BH'C$, since they have equal bases and the same altitude from C to ABH'. Therefore, as we have shown that triangle BHG is equal in area to triangle ABC, we can use the same procedure to show that each of the shaded triangles has an area equal to that of triangle ABC.

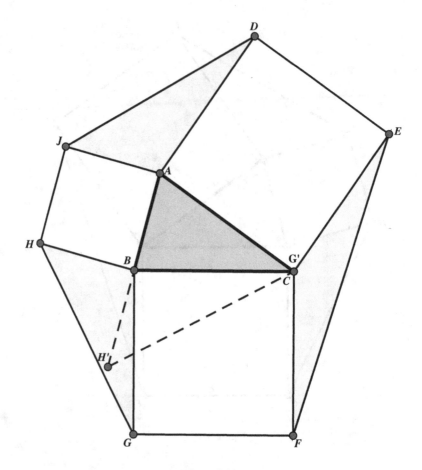

Figure 3.42

Adopting a Different Point of View

Problem: Comparing triangle sizes often presents challenging problems. Here we have right triangle *AFB* placed on square *ABCD* so that the hypotenuse is coincident with the side of the square, as shown in Figure 3.43. From the center of the square, point *P*, a line is drawn to the right-angle vertex *F*. What is the relationship between ∠*AFP* and ∠*BFP*?

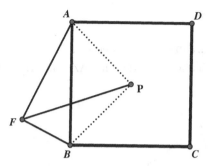

Figure 3.43

Solution: Since point P is the point of intersection of the diagonals, we draw them and find that $\angle APB$ is also a right angle. Looking at this problem from another point of view, we focus on quadrilateral $APBF$, which has supplementary opposite angles at vertices P and F. This makes it a cyclic quadrilateral and allows us to circumscribe a circle about it, as we show in Figure 3.44. Since $AP = BP$, we also have arc AP = arc BP, and therefore, $\angle AFP = \angle BFP$, which answers our question.

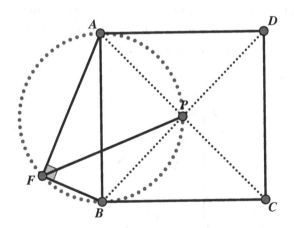

Figure 3.44

Discovering a Pattern

Problem: Here we are in search of a point P inside triangle ABC, shown in Figure 3.45, that will allow the triangles $\triangle APB$, $\triangle BPC$, and $\triangle CPA$ to have equal areas.

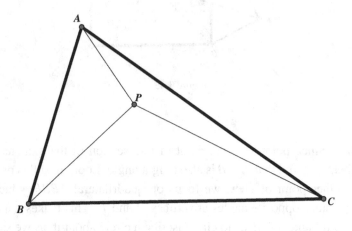

Figure 3.45

Solution: One needs to determine how a triangle can be divided into two equal triangles. This can be done by drawing the median, where in Figure 3.46 triangles ABM and ACM have equal bases, $BM = CM$, and the same altitude AD so that their areas are equal.

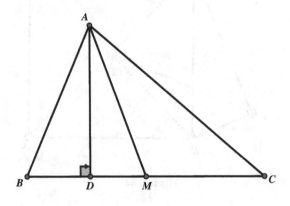

Figure 3.46

Having established that the median of a triangle divides the triangle into two equal areas, we can apply this as follows (see Figure 3.47): since area ($\triangle ABM$) = area($\triangle ACM$), and in triangle BPC, area ($\triangle BPM$) = area($\triangle CPM$), which by subtraction indicates that area ($\triangle APB$) = area($\triangle APC$). This can be repeated at any other portion of triangle ABC so that we will find area ($\triangle APB$) = area($\triangle APC$) = area($\triangle BPC$). Thus, the point P that partitions the original triangle into three equal-area triangles is the point of intersection of the medians and is known as the centroid of the triangle.

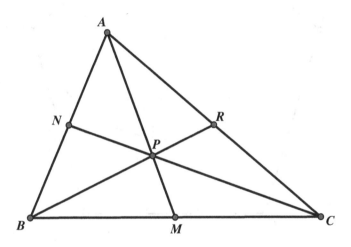

Figure 3.47

Working Backwards

Problem: A circle O is drawn so that it is tangent to sides BC and ED of a regular pentagon $ABCDE$, as shown in Figure 3.48. The challenge is to find the measure of arc BE of the circle with center O.

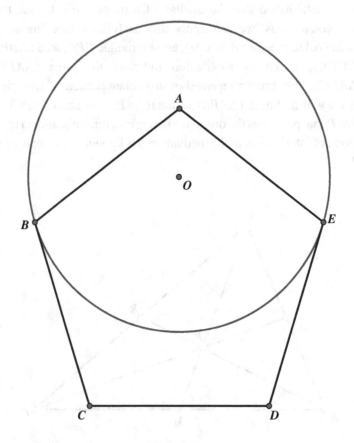

Figure 3.48

Solution: Working backwards, it would be good to be able to determine the measure of angle *BOE*, shown in Figure 3.49, as that would determine the measure of arc *BE*. At this point, the only angle measure we have available is the measure of any of the interior pentagon angles which are each equal to 108°. Since the radii to the point of tangency are perpendicular to the tangents, we have $OE \perp DE$.

Line *AO* extended is perpendicular to *CD* at point *F* and also bisects angle *ABE* so that $\angle EAO = \frac{1}{2}108° = 54°$. Furthermore, $\angle AEO = 108° - 90° = 18°$. Therefore, the exterior angle of $\triangle AOE$, namely, $\angle EOF = 54° + 18° = 72°$ and arc $BME = 2 \times 72° = 144°$.

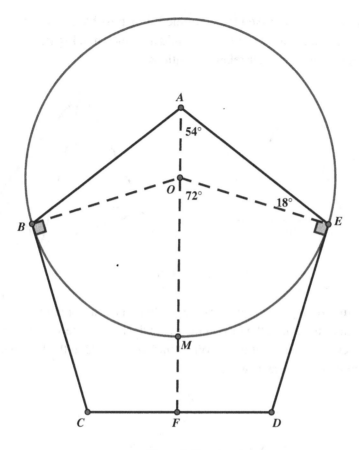

Figure 3.49

Perhaps, an easier way to get the angle measure of ∠*EOF* is to note that quadrilateral *EOFD*, whose opposite angles at vertices *E* and *F* are right angles, is a cyclic quadrilateral and, therefore, since ∠*EDF* = 108°, then its supplement ∠*EOF* = 72°. And thus, ∠*EOB* = 144°, which is then the measure of arc *BME*.

Intelligent Guessing and Testing

Problem: We are given the triangle shown in Figure 3.50 where the sides are marked with three colors, which in clockwise order are red, blue, and yellow. Here we are asked to determine if it is possible to use this triangle in whatever position to form the pyramid also shown in Figure 3.50 so that the colors remain in their relative positions.

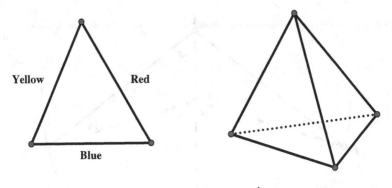

Figure 3.50

Solution: We show in Figure 3.51 the solution. It requires intelligent testing, specifically regarding the face in the rear, which has to be viewed from the back view in order for it to be correct and the triangle at the base needs to be seen in the reverse as well.

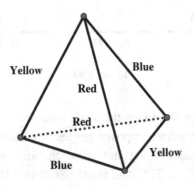

Figure 3.51

Problem: Here we are faced with the problem of dividing a circle into four equal parts. To do this symmetrically provides lots of opportunities such as those shown in Figure 3.52. However, to find a way to partition a circle into four equal parts which are *not* symmetric is the challenge here.

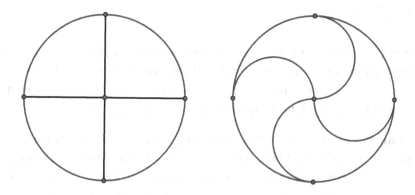

Figure 3.52

Solution: Draw semicircles on the diameter of the circle with diameters $\frac{1}{4}$, $\frac{1}{2}$, and $\frac{3}{4}$ of the large circle's diameter so that we get a partitioning, as is shown in Figure 3.53, where each region is one-quarter of the area of the original circle. Unlike the partitioning in Figure 3.52, these are not entirely symmetric. They are partly symmetric, namely, the outer two regions are symmetric to each other, and the two inner regions are also symmetric.

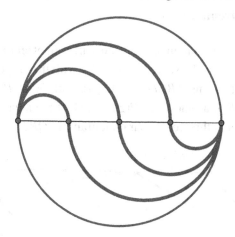

Figure 3.53

Suppose that the diameter of the big circle is 8. Then the area of one of the two symmetric inner regions is $\frac{\pi}{2}\left(2^2 - 1^1 + 3^2 - 2^2\right) = 4\pi$, and the area of one of the two symmetric outer regions is $\frac{\pi}{2}\left(1^2 + 4^2 - 3^2\right) = 4\pi$, thus, all four regions have the same area.

Organizing Data

Problem: Suppose Lisa and David have two pots of coins. Lisa says to David: "If you give me one coin, I will have twice as many coins as you have." David says to Lisa: "If you give me one coin, I will have as many coins as you have." How many coins does each of them have?

Solution: This appears as an elementary problem, especially when compared to others offered here. However, this problem is being included to demonstrate how using very elementary algebra can make a possibly confusing problem trivial. To simplify matters, we will let x equal the number of coins that David has, and let y equal the number of coins that Lisa has. Therefore, in the first case, we have $x - 1 = y + 1$. And in the second case, we have $x + 1 = 2(y - 1)$. When we solve these equations simultaneously, we get $y = 5$ and $x = 7$. We can check this by noting that when David gives one coin to Lisa, they will both have 6 coins and when Lisa gives one coin to David, he will have twice as many coins as Lisa.

Considering Extreme Cases

Problem: Here is a challenge from an early nineteenth century British geometry book that can be solved very easily with the proper technique. We are given parallelograms *ABCD* and *APQR*, with point *P* on side *BC* and point *D* on side *RQ*, as shown in Figure 3.54. If the area of parallelogram *ABCD* is 18, what is the area of parallelogram *APQR*?

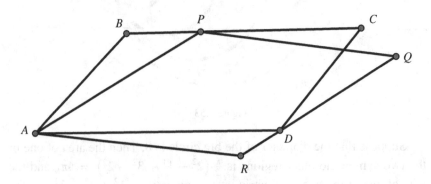

Figure 3.54

A common solution

This is by no means an easy problem to solve. First attempts to solve the problem would be to look for congruent relationships that would lead to equal areas. This method will lead nowhere. A clever method, although rather "off the beaten path," is to draw the line segment *PD*, as shown in Figure 3.55.

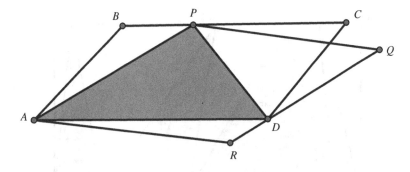

Figure 3.55

Then note that triangle *APD* can be shown to be one-half the area of each of the two triangles, since in each case it shares a base with each of the parallelograms as well as the related height. Although this is a rather clever approach to a challenging problem, there is yet an even shorter way to approach this problem, but this approach does not explain/prove that the position of *P* is indeed not relevant; it simply takes this for granted.

A cleverer solution

When the problem was posed, we were merely told that point *P* was on side *BC* but not where along the side it was to be placed. We can consider an extreme case. Therefore, we could have placed *P* to overlap point *B*. Similarly, point *D*, which was to be placed on side *RQ*, could just as easily have been placed to overlap point *R*. Under these circumstances, this would certainly fit the original problem's statement; the two parallelograms would overlap and consequently would have the same area. Therefore, the area of parallelogram *APQR* is 18.

Adopting a Different Point of View

Problem: In Figure 3.56, the dimensions of rectangle *AEFK* are given as width *AK* = 8, while the length *AE* is divided into four segments such that *AB* = 1, *BC* = 6, *CD* = 4, and *DE* = 2. This problem will allow us to appreciate the cleverness of an alternate way to determine what the area of the four shaded triangles is.

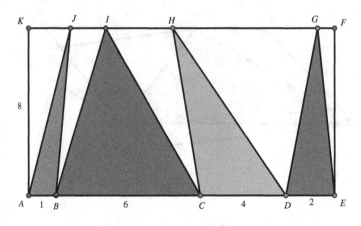

Figure 3.56

A common approach

The obvious approach is to find the area of each of the four triangles and get their sum. In all four cases, the altitude of the triangle equals the length of *AK* = 8. Therefore, the areas of the four triangles are as follows:

$$\triangle ABJ = \frac{1}{2} \times 1 \times 8 = 4$$

$$\triangle BCI = \frac{1}{2} \times 6 \times 8 = 24$$

$$\triangle CDH = \frac{1}{2} \times 4 \times 8 = 16$$

$$\triangle DEG = \frac{1}{2} \times 2 \times 8 = 8$$

The sum of these areas is $4 + 24 + 16 + 8 = 52$ square units.

A cleverer solution

We can make use of our strategy of adopting a different point of view to solve this problem. The triangles each have the same altitude, namely, 8. The sum of the bases of the four triangles equals the length of the longer side of the rectangle, which is 13. Thus, the area of the four shaded triangles is half the area of the rectangle or $\frac{1}{2} \times 13 \times 8 = 52$.

Working Backwards

Problem: A game can also demonstrate the power of mathematics. Max, Sam, and Jack are playing an unusual card game. In this game, when a player loses, he gives each of the other players as much money as they have. Max loses the first game and gives Sam and Jack as much money as they each have. Sam loses the second game and gives Max and Jack as much money as they each have. Jack loses the third game and gives Max and Sam as much money as they each have. They then decide to stop playing and each has exactly $8.00. How much money did each of them start with?

A common approach

The problem suggests we set up a series of equations designed to represent each game. We shall begin by representing the starting money for each as follows:

Max starts with x, Sam starts with y, and Jack starts with z.

Game #	Max	Sam	Jack
1	$x - y - z$	$2y$	$2z$
2	$2x - 2y - 2z$	$3y - x - z$	$4z$
3	$4x - 4y - 4z$	$6y - 2x - 2z$	$7z - x - y$

From the last transaction, we find that each of these values is 8. This gives us the following three equations in three variables:

$$4x - 4y - 4z = 8 \quad \text{or} \quad x - y - z = 2$$

$$-2x + 6y - 2z = 8 \quad \text{or} \quad -x + 3y - z = 4$$

$$-x - y + 7z = 8 \quad \text{or} \quad -x - y + 7z = 8$$

When we solve these three equations simultaneously, we obtain $x = 13$, $y = 7$, and $z = 4$.

A cleverer solution

Note that the problem gave the end situation and asked for the starting situation. This might give us a clue of a problem that will usually benefit from a strategy of working backwards. Note, too, that the statement of the situation shows that the same amount of money (namely, $3 \times \$8 = \$24$) is always "in play." Working backwards should provide an elegant solution.

	Max	Sam	Jack	Total
Game # 3	8	8	8	24
Game # 2	4	4	16	24
Game # 1	2	14	8	24
Start	13	7	4	24

Max started with $13, Sam started with $7, and Jack started with $4, the same answers as before but found in a more elegant fashion.

Considering Extreme Cases

Problem: Here is a popular mathematics question: A car is driving along a highway at a constant speed of 55 miles per hour. The driver notices a second car, exactly $\frac{1}{2}$ mile behind him. The second car passes the first car exactly 1 minute later. How fast was the second car traveling, assuming their speeds are constant?

A common approach

The traditional solution is to set up a series of "Rate × Time = Distance boxes," which has been the popular method to approach this sort of problem. This would be done as follows:

Rate × Time = Distance		
55	$\frac{1}{60}$	$\frac{55}{60}$
x	$\frac{1}{60}$	$\frac{x}{60}$

$$\frac{55}{60} + \frac{1}{2} = \frac{x}{60}, \quad \text{and} \quad x = 85$$

The second car was traveling at a speed of 85 miles per hour.

A cleverer solution

An alternate approach would be using the strategy of considering extremes. We assume that the first car is going *extremely* slowly, that is, at 0 miles per hour. Under these conditions, the second car travels $\frac{1}{2}$ mile in one minute to catch the first car. Thus, the second car would have to travel 30 miles per hour. When the first car is moving at 0 miles per hour, then the second car is traveling 30 mph faster than the first car. If, on the other hand, the first car is traveling at 55 miles per hour, then the second car must be traveling at 85 miles per hour (within the legal limit, of course!). Also, without thinking of extreme cases, one could argue as follows: With its *additional* velocity, the faster car needs 1 minute for the *additional* distance of $\frac{1}{2}$ mile, thus, the additional velocity must be 30 mph.

Considering Extreme Cases

Problem: We have two one-liter bottles. One contains a half liter of red wine and the other contains a half liter of white wine. We take a tablespoonful of the red wine and pour it into the white wine bottle and thoroughly mix the two-colored wines. Then we take a tablespoon of this new mixture (red wine and white wine) and pour it into the red wine bottle.

Is there more red wine in the white wine bottle, or more white wine in the red wine bottle?

A *common solution*

There are several common approaches, where problem solvers attempt to solve the problem using the given information, namely, the tablespoon. With some luck and cleverness, a correct solution may evolve, but it will not be easy and often not convincing.

A *cleverer solution*

We can see that the size of the spoon does not really matter, since there are large and small tablespoons: Suppose we use a very large tablespoon, one that is enormously large and actually can hold half liter of liquid — this would be an extreme consideration. When we pour the half liter of the red wine into the white wine bottle, the mixture is then 50% red wine and 50% white wine. After mixing these two together, we take our half-liter spoon and take one half quantity of this red wine–white wine mixture and pour it back into the red wine bottle. The mixture is now the same in both bottles; so, to answer our question, we can conclude that there is as much red wine in the white wine bottle as there is white wine in the red wine bottle.

Another way of approaching this problem is as follows: The missing red wine in the red wine bottle must be in the white wine bottle and vice versa, and since the volume of the liquid in both bottles is equal (half liter), these two parts (red wine in the white wine bottle and white wine in the red wine bottle) must be equal. This is further evidence why the size of the spoon is not important.

Making a Visual Representation

Problem: Time for a cute problem. If, on the average, a hen and a half can lay an egg and a half in a day and a half, how many eggs should six hens lay in eight days?

A *common solution*

This is an old problem that has survived the test of time. Traditionally, the problem is solved as follows: Since $\frac{3}{2}$ hens work for $\frac{3}{2}$ days, we may speak of the job of laying an egg and a half $\left(\frac{3}{2}\text{ eggs}\right)$ as taking $\left(\frac{3}{2}\right)\left(\frac{3}{2}\right)$ or $\left(\frac{9}{4}\right)$

"hen-days." Similarly, the second job takes 6×8 or 48 "hen-days". Thus, we form the following proportion:

Let x = the number of eggs laid by 6 hens in 8 days.
Then $\frac{\frac{9}{4} \text{ hen-days}}{48 \text{ hen-days}} = \frac{\frac{3}{2} \text{ eggs}}{x \text{ eggs}}$.

Multiplying the product of the means and extremes, we get $\left(\frac{9}{4}\right)(x) = 48\left(\frac{3}{2}\right)$ and $x = 32$.

A cleverer solution

However, as an alternate solution, we may set up the following visual representation (here in the form of a tabular layout) of the situation: $\frac{3}{2}$ hens lay $\frac{3}{2}$ eggs in $\frac{3}{2}$ days.

Double the number of hens:	**3 hens lay 3 eggs** in $\frac{3}{2}$ days
Double the number of days:	3 hens lay **6 eggs in 3 days**
One-third of the number of days:	3 hens lay **2 eggs in 1 day**
Double the number of hens:	**6 hens lay 4 eggs** in 1 day
Eight times the number of days:	6 hens lay **32 eggs in 8 days**

Therefore, 6 hens should lay 32 eggs in 8 days.

Solving a Problem from a Different Point of View

Problem: Consider two trains serving the Pittsburgh to New York route, a distance of 400 miles, starting toward each other at the same time (along the same tracks). One train is traveling uniformly at 60 mph and the other at 40 mph. At the same time, a bee begins to fly from the front of one of the trains, at a speed of 80 mph, toward the oncoming train. After touching the front of this second train, the bee immediately — without losing any time — reverses direction and flies toward the first train (still at the same speed of 80 mph). The bee continues this back-and-forth flying until the two trains collide, crushing the bee. The challenge is to determine how many miles the bee had flown.

Solution: One is naturally drawn to find the individual distances that the bee traveled. An immediate reaction is to set up an equation based on the relationship $\underbrace{\text{speed} \times \text{time}}_{\text{velocity}} = \text{distance}$. However, this back-and-forth path is rather difficult to determine because it requires considerable calculation. Even then, it is very difficult to solve the problem in this way.

A cleverer solution

A much more elegant approach would be to solve the problem from a different point of view. We seek the distance the bee traveled. If we knew the time the bee traveled, we could determine the bee's distance because we already know the bee's speed.

The length of time the bee traveled can be easily calculated because it traveled the entire time the two trains traveled (until they collided). To determine the time, t, the trains traveled, we set up an equation as follows: The distance of the first train is $60t$ and the distance of the second train is $40t$. The total distance the two trains traveled is 400 miles. Therefore, $60t + 40t = 400$ and $t = 4$, which is also the time the bee traveled. We can now find the distance the bee traveled, which is $(4)(80) = 320$ miles.

Considering Extreme Cases

Problem: Consider the globe of the earth with a rope wrapped tightly around the equator. The rope will be about 24,900 miles long. We now lengthen the rope by exactly 1 yard. We position this (now loose) rope around the equator so that it is uniformly spaced off the globe. The challenge is to determine if a mouse can fit under the rope (Figure 3.57).

Figure 3.57

Solution: The traditional way to determine the distance between the circumferences is to find the difference between the radii. Let R be the length of the radius of the circle formed by the rope (circumference $C + 1$) and r the length of the radius of the circle formed by the earth (circumference C), as shown in Figure 3.58.

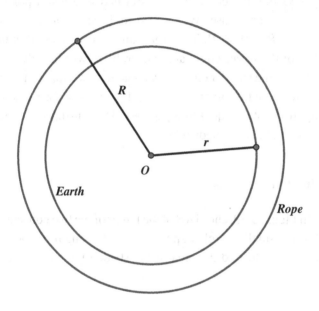

Figure 3.58

The familiar circumference formulas give us the following: $C = 2\pi r$, or $r = \frac{C}{2\pi}$, and we have $C + 1 = 2\pi R$, or $R = \frac{C+1}{2\pi}$. We need to find the difference of the radii, which is $R - r = \frac{C+1}{2\pi} - \frac{C}{2\pi} = \frac{1}{2\pi} \approx 0.159$ yards $\approx$ 5.7 inches. That is, there is a space of over $5\frac{1}{2}$ inches for a mouse to crawl under. Imagine, by lengthening the 24,900-mile rope by 1 yard, it lifted off the equator by about $5\frac{1}{2}$ inches!

A cleverer solution

Now for an even more elegant solution, which will consider the problem by using an extreme case — without loss of generality.

Keep in mind that the solution to the problem is independent of the circumference of the earth, since the end result did not include the circumference in the calculation. It only required calculating $\frac{1}{2\pi}$.

Therefore, using an extreme case, we may suppose the inner circle in Figure 3.58 is very small, so small that it has a zero-length radius, which essentially says that the circle has transformed into just a point. This is, obviously, the extreme case. We seek to find the difference between the radii, $R - r = R - 0 = R$. Therefore, all we need to find is the length of the radius of the larger circle and our problem will be solved. With the circumference of the smaller circle now equal to 0, we apply the formula for the circumference of the larger circle: $C + 1 = 0 + 1 = 2\pi R$, then $R = \frac{1}{2\pi}$, which is approximately equal to $5\frac{1}{2}$ inches. Thus, using the extreme case we solved the problem immediately.

Considering Extreme Cases

Problem: In Figure 3.59, chord AB of the larger of the two concentric circles is tangent to the smaller circle at point T. As the length of chord $AB = 8$, We are challenged to find the area of the shaded region between the two circles.

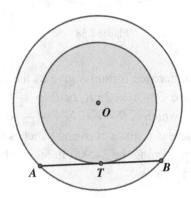

Figure 3.59

Solution: We know that a radius is perpendicular to a tangent at the point of contact (T). Furthermore, a radius perpendicular to a chord divides the chord into two equal segments. Thus, in Figure 3.60, $AT = BT = 4$. We know that the area of the region between the two circles (the doughnut shape) can be found by obtaining the difference between the areas of the two circles.

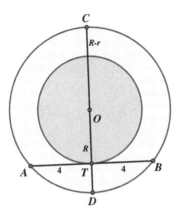

Figure 3.60

Thus, the area of this region between the two circles equals $\pi R^2 - \pi r^2 = \pi(R^2 - r^2)$. Now, $OC = R$ and $OT = r$, $CT = (R+r)$ and $DT = (R-r)$, which can be seen in Figure 3.60. Recall that the product of the segments of two intersecting chords of a circle is equal. Therefore, we obtain $(R-r)(R+r) = 4 \times 4$, and then $R^2 - r^2 = 16$. Thus, the area of the region between the two circles equals 16π square units.

We could have also solved this problem by drawing line segment OA. We then created a right triangle ATO, in which $(OA)^2 = (OT)^2 + 4^2$ or $R^2 - r^2 = 16$, which again gives us the area as 16π square units.

A cleverer solution

We can also look at this problem by considering an extreme case. Let's assume that the smaller circle gets smaller and smaller, until it becomes a point that coincides with point O. Then, AB becomes a diameter of the larger circle, and the area of the region between the two circles becomes the area of the larger circle, which equals $\pi R^2 = 16\pi$.

chord into two equal segments. Thus, in Figure 3.27, $AC = BC$, and
We know that the area of the region between the two circles (the doughnut
shape) can be found by subtracting the difference between the areas of the
two circles.

Thus the area of this region is $\pi w = \pi (R + r)^2 - \pi r$ equals
$R^2 + 2rw = \pi (R^2 + 2rw) - \pi R$, and $(R^2 + 2rw) = \pi (R + r)^2$ and
$(R^2 + rR)$, which can be written in Figure 3.27. Recall that the product
of the lengths of two intersecting chords of a circle is equal. Therefore,
and $R^2 = (R \times r)$, $R \times R = r$, and then $R - r$ equals 1. Thus, the area
of the region between the two circles equals 1π square unit.

We could have solved this problem by drawing the segment CA.
We then created a right triangle ACB in which $(CA)^2 = (BC)^2 + r^2$ or
$R^2 = r^2 + 1$, from which again gives us the area as 1π square units.

A Closer Look

We find also in this problem by considering an extreme case, let's
imagine that the smaller circle gets smaller and smaller until it becomes a
point, now coincides with point O. Then AB becomes a diameter of the
larger circle, and the area of the region between the two circles becomes
the area of the larger circle which equals $R^2 = 1\pi$.

Chapter 4

Algebra: A Unique Problem-Solving Tool

When thinking of the basics of mathematics, the topic of algebra follows immediately after arithmetic. Although algebra is often considered a powerful tool to solve problems, there are many problems where the use of algebra can lead to unnecessarily complicated solutions. Before we exhibit numerous examples where algebra shows its power, we will consider a few problems whose solutions are complicated by using algebra techniques. In this chapter you will experience some of the mistakes people can make with algebra, which then, of course, lead to absurd results. Your journey through this chapter should give you a good impression of the power and usefulness of algebra as well as the issues one should be aware of.

A Problem with Marbles

Sandy has a bag containing marbles. She gives half of them to Bernie and then one-third of the remaining marbles to Peter. Finally, she has 6 marbles left. How many marbles were in Sandy's bag initially?

If one denotes the unknown number of marbles with x and establishes an equation following the text, the result would be $x - \frac{x}{2} - \frac{1}{3}\left(x - \frac{x}{2}\right) = 6$. One could argue that solving this equation leaves one open to making trivial mistakes. Therefore, by working backwards and making a sketch, such as a circle diagram as shown in Figure 4.1, it should be not difficult to see that the right half of the circle must have 9 marbles, and thus, the whole circle (which corresponds to Sandy's initial bag) must have contained 18 marbles.

Figure 4.1

In the following problem, we have an example where "algebra" is only one powerful tool among others (here "using illustrative figures").

Sum of Three or More Consecutive Natural Numbers

Here we can experience algebra as a short and instructive tool for visualizing the sum of three consecutive natural numbers. We can write this sum as $(n-1) + n + (n+1)$; when simplified, we get $3n$, which indicates that such a sum is always divisible by 3. The same is possible for $5, 7, 9, \ldots, 2k+1$, in fact for every odd number of consecutive natural numbers. Consider the situation for 7 consecutive numbers:

$$(n-3) + (n-2) + (n-1) + n + (n+1) + (n+2) + (n+3) = 7n.$$

Also, the sum of an even number of consecutive natural numbers can be well understood by using algebra. What can be said about the divisibility of the sum of an even number of consecutive natural numbers, such as 4 or 6 consecutive numbers?

Sum of 4 consecutive numbers is as follows:

$$(n-1) + n + (n+1) + (n+2) = 4n + 2 = 2(2n+1)$$

Sum of 6 consecutive numbers is as follows:

$$(n-2)+(n-1)+n+(n+1)+(n+2)+(n+3) = 6n+3 = 3(2n+1)$$

Sum of $2k$ consecutive numbers is as follows:

$$(n-(k-1))+(n-(k-2))+\cdots+n+\cdots+(n+(k-1))+(n+k)$$
$$= (2k)n + k = k(2n+1)$$

One can see, with the help of algebra, that it is never possible that a sum of an even number $(2k)$ of consecutive natural numbers is divisible by that even number $2k$, but this sum is always divisible by k. This is a famous, and rather elementary, example of how algebra can help explain numerical patterns with minimal effort.

An impressive alternative to explaining these concepts is done by using illustrative figures.

In case of an odd number $2k+1$ of consecutive natural numbers, there is always a mean value, say m. Then one can imagine to cut off the part of numbers greater than m and move this part to the left yielding a rectangle with a width $2k+1$ and a height m, and it can be seen visibly that such a sum is divisible by $2k+1$ (see Figure 4.2).

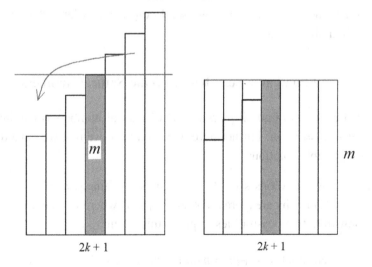

Figure 4.2

In case of an even number $2k$ of consecutive natural numbers, one can take the right part and put it onto the left part yielding a rectangle with width k, which demonstrates visibly that such a sum is always divisible by k (see Figure 4.3).

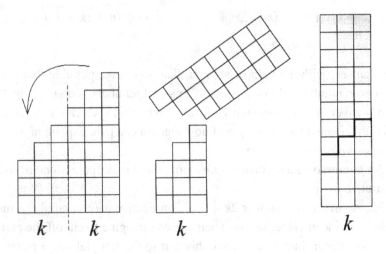

Figure 4.3

Now that we've experienced a comparison of algebraic solutions to logic techniques and visual methods, we shall consider problems that are best approached algebraically.

An Unusual Equation in Search for Specific Natural Numbers

The challenge is to find all pairs (n, m) of natural numbers (including 0) where $n + m = n \cdot m$. Furthermore, we need to prove that we have found all such possible solutions.

Solution: Two solutions seem to arise quickly. Namely, $n = 0 = m$ and $n = 2 = m$. But why are there no others? Here is where we can use algebra to demonstrate two possibilities to prove this result.

(1) The given equation $n + m = n \cdot m$ is equivalent to $1 = n \cdot m - n - m + 1$, and further this is equivalent to $1 = (n - 1)(m - 1)$. And therefore,

the only two ways that 1 is a product of two integers are $1 = 1 \cdot 1$ and $1 = (-1) \cdot (-1)$. In the first case, we would have $n = 2 = m$ and in the second case, we would have $n = 0 = m$.

(2) For $n = 0$, we have $m = 0$, and it is clear that $n = 1$ can never be a solution since that would mean $1 + m = m$. So, we can assume for what follows $n \geq 2$:

$$n + m = n \cdot m \Leftrightarrow n = m \cdot (n - 1) \Leftrightarrow \frac{n}{n-1} = m \Leftrightarrow 1 + \frac{1}{n-1} = m,$$

the only way that (for $n \geq 2$) $\frac{1}{n-1}$ is an integer is $n = 2$ which yields $m = 2$.

Sum of Squares Equals Sum of More Squares

Here is one problem that sounds very simple and yet requires a little bit of thought. Begin by taking the sum of any three squares and multiplying it by 3. We now need to find four squares that will have the same sum. For example, $3(2^2 + 3^2 + 4^2) = 87 = 9^2 + 2^2 + 1^2 + 1^2$. Or perhaps as another example: $3(2^2 + 3^2 + 3^2) = 66 = 5^2 + 4^2 + 4^2 + 3^2$. This can be sometimes rather frustrating and yet also delightfully challenging when successful. At this point, we might want to see a justification so that we are not left with an unsolvable situation. We provide a simple algebraic proof:

$$3(a^2 + b^2 + c^2) = (a + b + c)^2 + (b^2 - 2bc + c^2)$$
$$+ (c^2 - 2ca + a^2) + (a^2 - 2ab + b^2)$$
$$= (a + b + c)^2 + (b - c)^2 + (c - a)^2 + (a - b)^2$$

Using this relationship will provide other possible solutions.

Arithmagons with Four Vertices

Let us first define an arithmagon, which is a polygon-shaped figure where there is an encircled number at each vertex and a boxed number on each side so that the boxed number is the sum of the two circled numbers between which it is located.

Now it is very easy to find the numbers on the sides (in squares) given the numbers in the circles (at vertices), but the other way round, it could be more interesting. That is, given the numbers in the squares on the edges (a, b, c, d), shown in Figure 4.4, we find possible numbers for the circles at the vertices (all these numbers should be natural numbers).

(a) Are there solutions in the following four-sided arithmagons shown in Figure 4.4? If yes, how many? If no, then why?
(b) Give necessary and sufficient conditions for $a, b, c, d \in \mathbb{N}$ so that there are solutions! How many such conditions do exist, if any exist at all?

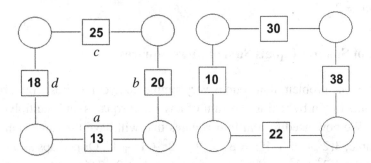

Figure 4.4

Solution: For the first arithmagon, we can use trial and error to easily find a solution. Let us consider the bottom-left vertex (circle) with the value 7 (Figure 4.5). Then at the bottom-right circle, there must be 6 so that the sum of 13 is on the bottom side. On the top-right vertex, there must be 14 in order to justify the sum of 20 on the right-side square. Finally, on the top-left vertex, we have 11 so that we can have the sum of 25 on the top side, and this fits as well to the sum 18 on the left side.

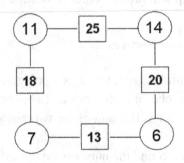

Figure 4.5

Had we started at the bottom-left vertex with the number other than 7, that is, using any other natural number of $\{1, 2, \ldots, 12\}$, we also would have got a solution. Suppose we had taken 3, then we would have arrived at another solution as shown in Figure 4.6.

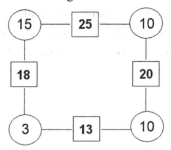

Figure 4.6

There must be a solution also with any number from the set $\{1, 2, \ldots, 12\}$ at the bottom left because as the value of the number at the bottom-left vertex gets smaller (or larger), the number at the top-left vertex gets bigger (or smaller) by the same amount. Thus, at the end, it must fit with the same sum of 18. So, altogether we have 12 solutions, and if we count 0 among the natural numbers, the number on the bottom-left vertex could be one of these numbers $\{0, 1, \ldots, 12, 13\}$, thereby, resulting in 14 solutions. In general, the arithmagon will have solutions if we let a be the minimum (smallest value) of $\{a, b, c, d\}$ then the number of solutions is $a - 1$ or $a + 1$, depending on whether or not one counts 0 among the natural numbers.

When we try the same technique for the second arithmagon, we fail by trying 7 again at the bottom-left vertex, as shown in Figure 4.7.

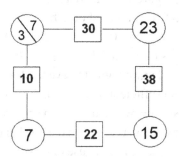

Figure 4.7

At the top-left vertex, on the one hand, we should write 3 to get a sum of 10 on the left edge, and on the other hand, we should write 7 so as to get a sum of 30 on the top edge. A conflict arises. But also any other number $7 + x$ at the bottom-left vertex instead of 7 does not work (see Figure 4.8): At the top-left vertex, on the one hand, we should write $3 - x$ to get a sum of 10 on the left edge, and on the other hand, we should write $7 - x$ so as to get a sum of 30 on the top edge, again a contradiction!

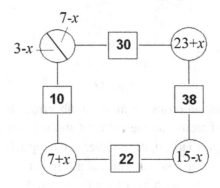

Figure 4.8

Our task would be to determine if there is a solution by merely inspecting a, b, c, d. Since the sum of two natural numbers is at least 2 (not counting 0 as part of the natural numbers), we have each of the numbers $a, b, c, d \geq 2$ must hold (a necessary condition) for a solvable four-sided arithmagon. Furthermore, considering Figure 4.9, with a and c, two vertices are summed up. The sum of the bottom two vertices should be a, and the sum of the top two vertices should be c. Therefore, for $a + c$, we have the sum of the four vertices, and the same holds for $b + d$. Thus, $a + c = b + d$ is another necessary condition for a four-sided arithmagon to have a solution. And these two conditions together are also sufficient, because in this case one can always find appropriate numbers, starting with, say, x, at the bottom left vertex, and at the end at the top left vertex one gets $d - x$ on the one hand, and $a + c - b - x$ on the other hand, which is the same because of $a + c = b + d$, as we can see in Figure 4.9.

$$c - b + (a - x) = a + c - b - x = d - x$$

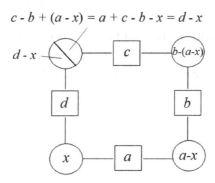

Figure 4.9

If one considers 0 to be included in the set of natural numbers, $\mathbb{N}$, then the condition $a, b, c, d \geq 2$ can be omitted for a solvable arithmagon because in this case, the minimum sum of two natural numbers is 0 and this does not yield a new condition.

Arithmagons with Three Vertices

There is a substantial difference between four-sided arithmagons and three-sided arithmagons, which have either no solution or exactly one solution. In case of a solvable four-sided arithmagon, it' is rather easy to find a solution, but three-sided arithmagons are not so easily solved. We use algebra for two purposes: First, to see how to arrive at a solution, if one exists, although, this could also be accomplished without algebra as well; second, to see under which conditions for $a, b, c \in \mathbb{N}$ there is a solution with natural numbers — here algebra is a powerful tool. We begin with the diagram shown in Figure 4.10.

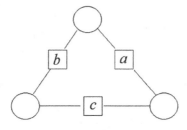

Figure 4.10

In Figure 4.11, we will begin with x at the bottom-left vertex, then at the bottom-right vertex there must be $c - x$ and the value on the top must be on the one hand $b - x$, and on the other hand $a - (c - x) = a - c + x$; these two numbers have to be equal. Therefore, $b - x = a - c + x$, which is equivalent to $x = \frac{b+c-a}{2}$.

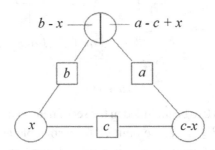

Figure 4.11

Therefore, we have $x = \frac{b+c-a}{2}$, which determines x and subsequently determines the two numbers at the other vertices. Furthermore, we can see that $b + c - a$ must be an *even* natural number so that the result will not be a fraction. That means $b + c > a$, and analogously $a + c > b$ and $a + b > c$. Either all three of $a, b, c \in \mathbb{N}$ are even or two of them are odd and one is even. If one includes 0 among the natural numbers, the conditions $b + c > a$, $a + c > b$, and $a + b > c$ change to $b + c \geq a$, $a + c \geq b$, and $a + b \geq c$.

For instance, the left three-sided arithmagon in Figure 4.12 is solvable in $\mathbb{N}$, as one can find the missing numbers also without algebra. The right arithmagon is not solvable in $\mathbb{N}$ (three odd numbers).

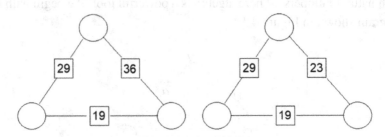

Figure 4.12

Interchanging × and + Without Changing the Value (I)

It is easy to see $6 \times \frac{6}{5} = 6 + \frac{6}{5}$, since $\frac{6 \times 6}{5} = \frac{36}{5}$ and $\frac{30}{5} + \frac{6}{5} = \frac{36}{5}$.

Are there other such examples? If so, then we need to find all of them and prove we have them all.

Solution: Here is an example where algebra becomes essential. The first guess could be that it works with numbers of the form $n \times \frac{n}{n-1} = n + \frac{n}{n-1}$ with $n > 1$. Actually, this is true and can be seen by applying algebra to this equation by multiplying both sides by $(n - 1)$, to get $n^2 = n^2 - n + n$, which implies that the equality above is true for every such n. However, the question still remains: Are there other such examples?

Let us use three different variables. Once again, we begin with $a \times \frac{b}{c} = a + \frac{b}{c}$, where from the beginning we have the condition $c \neq 0$. In case of $a = 0 = b$, the value of $c \neq 0$ could be chosen arbitrarily. The above equation can be written equivalently as $(a - 1) \times \frac{b}{c} = a$. Obviously, for $a = 1$, this is not possible, since 0 is not equal to 1. For $a \neq 1$, we get $\frac{b}{c} = \frac{a}{a-1}$. Here we see, the reduced version of $\frac{b}{c}$ must be $\frac{a}{a-1}$, so essentially there are no other such examples as we had above, namely, $n \times \frac{n}{n-1} = n + \frac{n}{n-1}$.

Interchanging × and + Without Changing the Value (II)

It is easy to see $6 \times 1.2 = 6 + 1.2$, since both sides are equal to 7.2.

Are there other such examples? If yes, we need to find all of them and then prove all have been identified.

Solution: This is very similar to the previous problem; the difference is that in the previous problem, we had (non-negative) fractions, and here we have real numbers that can be positive or negative. As we embark on the challenge presented above, we have to solve the equation $x \cdot y = x + y$ and not be limited to integers. First, we see that $x = 1$ is impossible because then $y = 1 + y$ is a contradiction. And for $x \neq 1$, we can say $x \cdot y = x + y \Leftrightarrow y \cdot (x - 1) = x \overset{x \neq 1}{\Leftrightarrow} y = \frac{x}{x-1} = \frac{1}{x-1} + 1$ and, thus, we see that all points (x, y) lie on the graph of the function $y = \frac{1}{x-1} + 1$ which is a *hyperbola* (see Figure 4.13).

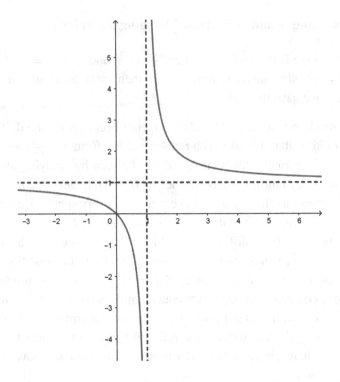

Figure 4.13

This is a generalization of the above problem (Interchanging $\times$ and $+$ without changing the value (I)) where we had the restrictions that x was a non-negative integer, moreover, the variables were n or a instead of x, and we had y as a positive rational number. (Note: Above we had $\frac{n}{n-1}$ or $\frac{b}{c}$.)

A Curiosity with Mixed Numbers

It is easy to see $\sqrt{5\frac{5}{24}} = 5\sqrt{\frac{5}{24}}$ (with a *mixed number* under the first square root). This can be equated as follows:

$$\sqrt{5\frac{5}{24}} = \sqrt{\frac{24 \cdot 5 + 5}{24}} = \sqrt{\frac{5 \cdot (24 + 1)}{24}} = \sqrt{\frac{125}{24}},$$

and the right side is $5\sqrt{\frac{5}{24}} = \sqrt{25} \cdot \sqrt{\frac{5}{24}} = \sqrt{\frac{25 \cdot 5}{24}} = \sqrt{\frac{125}{24}}.$

Is this the only such example of this relationship or are there others? If there are others, then our challenge is to find all of them for non-negative numbers.

Solution: Another example to consider would be $\sqrt{6\frac{6}{35}} = 6\sqrt{\frac{6}{35}}$. Are all possible examples built like $\sqrt{n\frac{n}{n^2-1}} = n\sqrt{\frac{n}{n^2-1}}$? Algebra can help us to have an overview over all possible cases.

Algebraically, we need to solve the equation $\sqrt{x\frac{y}{z}} = x\sqrt{\frac{y}{z}}$, where z is a positive integer and x, y are non-negative integers. For $x = 0$, we have $\sqrt{\frac{y}{z}} = 0$ with the solution $y = 0$ and $z > 0$ an arbitrary positive integer. For $x > 0$, when we square both sides of $\sqrt{x\frac{y}{z}} = x\sqrt{\frac{y}{z}}$, we get $z = \frac{y(x+1)(x-1)}{x}$, and $x \neq 1$ or else $z = 0$, which is not acceptable here. Considering the greatest common divisor, we see $\gcd(x, x + 1) = 1 = \gcd(x, x - 1)$. Therefore, the only chance for $z \in \mathbb{N}$ is for x to be a divisor of y so that $y = kx$, $(k \in \mathbb{N})$. We, thus, have the following solutions:

(1) $x = 0 = y, z > 0$ arbitrary
(2) $x \neq 0; x \neq 1$ arbitrary, $y = k \cdot x (k \in \mathbb{N}), z = k \cdot (x^2 - 1)$

So essentially, there are no other interesting solutions than what we had above. Just extending the fraction with k is also not a new solution: $\sqrt{n\frac{n \cdot k}{(n^2-1)\cdot k}} = n\sqrt{\frac{n \cdot k}{(n^2-1)\cdot k}}$. Maybe one may have had the right conjecture earlier, but now, with the use of algebra, one *knows* what is happening.

A Surprise in a Sequence

We begin by taking two arbitrary initial *positive* numbers. Then with the following procedure, we create the *next* number: Add 1 to the second number and divide the result by the first number and then add 1 to that result and divide it by the second number. Let us do that with an example by taking initially the numbers $\frac{3}{4}$ and 2. Then the first step would yield $\frac{2+1}{\frac{3}{4}} = \frac{3}{\frac{3}{4}} = 4$, the second step would be $\frac{4+1}{2} = \frac{5}{2}$, and if one continues this process, the sequence will look like this: $\underbrace{\frac{3}{4}, 2}_{\substack{\text{arbitrarily}\\\text{chosen}}}, 4, \frac{5}{2}, \frac{7}{8}, \frac{3}{4}, 2, \ldots$

Here we can see that the sixth number in the sequence equals the first number (where we began) and the seventh number equals the second number. Take other initial numbers and this will hold true. Once again, it will be through algebra that we will explain this phenomenon.

Solution: One can take any numbers $a_1 \neq 0$ and $a_2 \neq 0$ as starting numbers; we need to show that they will always yield $a_6 = a_1$ and $a_7 = a_2$. With algebra, the proof involves just a few calculations with fractions and is very expedient. The calculation for a_4 goes like this: $a_4 = \dfrac{\frac{a_2+1}{a_1}+1}{a_2} = \dfrac{a_1+a_2+1}{a_1 a_2}$. The one for a_5 like this: $a_5 = \dfrac{\frac{a_1+a_2+1}{a_1 a_2}+1}{\frac{a_2+1}{a_1}} = \dfrac{a_1+a_2+1+a_1 a_2}{a_2(a_2+1)} = \dfrac{(a_1+1)(a_2+1)}{a_2(a_2+1)} = \dfrac{a_1+1}{a_2}$; then we get $a_6 = \dfrac{\frac{a_1+1}{a_2}+1}{\frac{a_1+a_2+1}{a_1 a_2}} = \dfrac{a_1+a_2+1}{\frac{a_1+a_2+1}{a_1}} = a_1$ and finally $a_7 = \dfrac{a_1+1}{\frac{a_1+1}{a_2}} = a_2$.

Altogether we can summarize:

$$\underbrace{a_1, a_2}_{\substack{\text{arbitrarily}\\\text{chosen}}}, \underbrace{\frac{a_2+1}{a_1}}_{a_3}, \underbrace{\frac{a_1+a_2+1}{a_1 a_2}}_{a_4}, \underbrace{\frac{a_1+1}{a_2}}_{a_5}, \underbrace{a_1}_{a_6}, \underbrace{a_2}_{a_7} \cdots$$

There is another interesting question for the motivated reader. Is it possible that the equality $a_n = a_1, a_{n+1} = a_2$ can hold true for some value of $n < 6$? We know from the above that it holds true at the latest for $n = 6$. Let us try $n = 2$: This would mean $a_2 = a_1$, $\underbrace{\frac{a_2+1}{a_1}}_{a_3} = a_2$, and this equation

system has the only positive solution $a_1 = a_2 = \frac{1+\sqrt{5}}{2} = \varphi \approx 1.618$, which happens to be the golden ratio. Explanation: The equation system yields $\frac{a_1+1}{a_1} = a_1$, which is equivalent with $a_1^2 - a_1 - 1 = 0$ and solving this with the well-known formula for quadratic equations yields the two solutions $a_1 = \frac{1}{2} \pm \sqrt{\frac{1}{4}+1}$; one of them is negative and the other one can be written as above and is the golden ratio. So, if we start with $a_1 = a_2 = \varphi$, we have a constant sequence.

What about $n = 3$? This would mean

$$\overbrace{\frac{\frac{a_2+1}{a_1}}{a_1}}^{a_3} = a_1$$

$$\underbrace{\frac{a_1+a_2+1}{a_1 a_2}}_{a_4} = a_2$$

and this equation system has, again, the only positive solution $a_1 = a_2 = \frac{1+\sqrt{5}}{2} = \varphi \approx 1.618$, and the same holds for $n = 4$ and for $n = 5$. Therefore, the only way that the equality $a_n = a_1, a_{n+1} = a_2$ happens for some $n < 6$ is the constant sequence in the case of $a_1 = a_2 = \frac{1+\sqrt{5}}{2} = \varphi \approx 1.618$.

Sum of Unit Fractions

In ancient Egypt, aside from the fraction $\frac{2}{3}$, all measurements were made by using unit fractions. Therefore, it was important to determine how fractional measurements can be expressed as the sum of various unit fractions. Our challenge here is to show that all unit fractions between 0 and 1 of the form $\frac{1}{n}$ ($n \geq 2$) and fractions of the form $\frac{2}{n}$ ($n \geq 3$) can be represented as sums of *different* unit fractions. (Note: The representation $\frac{2}{n} = \frac{1}{n} + \frac{1}{n}$ is not acceptable because the unit fractions are not different, as required.)

Solution: One can easily check some small numbers for n: $\frac{1}{2} = \frac{1}{3} + \frac{1}{6}$, $\frac{1}{3} = \frac{1}{4} + \frac{1}{12}$, $\frac{1}{4} = \frac{1}{6} + \frac{1}{12} = \frac{1}{5} + \frac{1}{20}$ .... Sometimes, one will find more than one such representation as a sum of different unit fractions as we have shown above with the fraction $\frac{1}{4}$. Yet, we are challenged to prove that there is always such a representation available. Suppose we start in the case of $\frac{1}{n}$ with the largest possible unit fraction as the first summand: $\frac{1}{n} = \frac{1}{n+1} + x$. And we are lucky, as x will turn out as a *different* unit fraction $x = \frac{1}{n} - \frac{1}{n+1} = \frac{n+1-n}{n(n+1)} = \frac{1}{n(n+1)}$. (Note: For $n \geq 2$, we have $\frac{1}{n(n+1)} < \frac{1}{n+1}$.) Thus, we have demonstrated what was requested.

In case of $\frac{2}{n}$ ($n \geq 3$), we can distinguish two cases. If, on the one hand, n is an even number, say, $n = 2k$, then $\frac{2}{n} = \frac{1}{k}$, and from this we know that there is such a representation. If on the other hand, n is an odd number, say, $n = 2k - 1$, then $\frac{2}{n} = \frac{2}{2k-1}$.

We use the same technique as above, that is, to try the biggest possible unit fraction. Here we take $\frac{1}{k}$, as the first summand, and we are successful as the remaining summand must then be a unit fraction as well: $\frac{2}{2k-1} - \frac{1}{k} = \frac{1}{k(2k-1)}$, and thus, $\frac{2}{2k-1} = \frac{1}{k} + \frac{1}{k(2k-1)}$.

A Number Pyramid with Fractions

In an arbitrarily big number pyramid, as shown in Figure 4.14, the left-most "bricks" are filled with unit fractions, beginning at the top with $\frac{1}{1}$. The interior bricks are filled following the well-known rule:

Each brick is the sum of the two bricks below it.

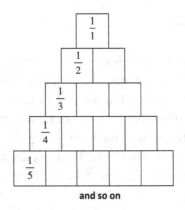

and so on

Figure 4.14

Then one can see two surprising phenomena:

(a) The whole number pyramid contains only *unit fractions* (after reducing the fractions).
(b) The number pyramid is *symmetric*.

Our challenge here is to justify why this is true.

Solution: If one fills the empty bricks above, one gets Figure 4.15.

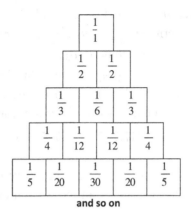

Figure 4.15

But such a filled-in number pyramid, even if it were 1,000 times larger, is not a proof. And for that purpose, we introduce variables, and once again an algebraic procedure is welcomed.

Let us number the oblique columns running from top left to bottom right with $1, \ldots, n, \ldots$ and the oblique columns running from top right to bottom left with $1, \ldots, m, \ldots$ and denote the numbers in each brick with $a_{m,n}$ as shown in Figure 4.16.

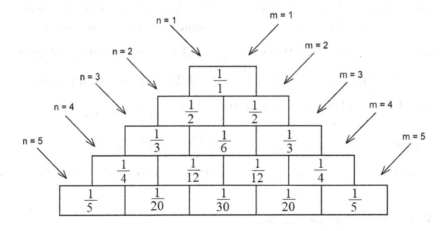

Figure 4.16

Then we know that $a_{1,n} = \frac{1}{n}$ (first given oblique column at the left). Then we can easily determine $a_{2,n} = a_{1,n} - a_{1,n+1} = \frac{1}{n} - \frac{1}{n+1} = \frac{1}{n(n+1)}$, again a unit fraction. Now we come to $a_{3,n} = a_{2,n} - a_{2,n+1} = \frac{1}{n(n+1)} - \frac{1}{(n+1)(n+2)} = \frac{n+2-n}{n(n+1)(n+2)} = \frac{2}{n(n+1)(n+2)}$, and since the denominator is a product of three consecutive natural numbers, it is surely divisible by 2, and we know that also $a_{3,n}$ is a unit fraction (after reducing it). Then we go to $a_{4,n} = a_{3,n} - a_{3,n+1} = \frac{2}{n(n+1)(n+2)} - \frac{2}{(n+1)(n+2)(n+3)} = \frac{2(n+3)-2n}{n(n+1)(n+2)(n+3)} = \frac{1\times2\times3}{n(n+1)(n+2)(n+3)}$.

With an analogous discussion to that above, we can be sure that the denominator is divisible by 2 and 3. In the general case, we get with the same technique:

$$a_{m,n} = \frac{(m-1)!}{\frac{(n+m-1)!}{(n-1)!}} = \frac{(m-1)!(n-1)!}{(n+m-1)!}$$

Here we used for abbreviation the factorial notation, which is $n! = n \times (n-1) \times (n-2) \times \cdots \times 2 \times 1$.

One could produce a proof using mathematical induction for m. This term is on the one hand symmetric in m and n, which proves the symmetry of the number pyramid, and on the other hand it can be written as $a_{m,n} = \frac{1}{\frac{(n+m-1)!}{(m-1)!(n-1)!}} = \frac{1}{n\binom{n+m-1}{m-1}}$, which proves that it is a unit fraction. Here we used the well-known notation of *binomial coefficients* which always represent natural numbers since they indicate a *number of possibilities*. In general, the binomial coefficient $\binom{n}{k}$ describes the number of possibilities to select k different elements out of $n \geq k$ different elements. It is standard knowledge of elementary combinatorics that $\binom{n}{k} = \frac{n!}{k!(n-k)!}$, which was used above. As an example, there are $\binom{10}{4} = \frac{10!}{4!\times6!} = \frac{10\times9\times8\times7}{4\times3\times2\times1} = 210$ possibilities to choose a team of 4 persons from a group of 10 persons.

Which Is the Larger Fraction?

There are countless curiosities in mathematics that at first sight generate amazement. Through the use of algebra, we can make some sense out of these surprising and counterintuitive phenomena. Consider the two fractions:

$$\text{(a)} \ \frac{59\sqrt{59} + 3\sqrt{3}}{59\sqrt{3} + 3\sqrt{59}} \qquad \text{(b)} \ \frac{59\sqrt{59} - 3\sqrt{3}}{59\sqrt{3} - 3\sqrt{59}}$$

The challenge is to determine which of these two fractions (a) or (b) is the greater. Furthermore, we need to justify our response, which is where algebra can be extremely useful.

Solution: The first issue to consider is if the numbers 59 and 3 are critical to this question. For the first approach, one will take a calculator or a computer and find the following values: $\frac{59\sqrt{59}+3\sqrt{3}}{59\sqrt{3}+3\sqrt{59}} \approx 3.660205$ and $\frac{59\sqrt{59}-3\sqrt{3}}{59\sqrt{3}-3\sqrt{59}} \approx 5.660205$.

Unexpectedly, we find that the second fraction is larger, despite the fact that it has the same numbers and involves minus signs.

However, to see if there is any pattern independent of the numbers, we shall use another pair of numbers, 45 and 5, with the same pattern to see if anything noteworthy results.

$\frac{43\sqrt{43}+5\sqrt{5}}{43\sqrt{5}+5\sqrt{43}} \approx 2.273573$ and $\frac{43\sqrt{43}-5\sqrt{5}}{43\sqrt{5}-5\sqrt{43}} \approx 4.273573$, which suggests the conjecture that (b) is always 2 greater than (a). But even if we had 10 or 100 more examples, we would not have a proof. Therefore, we have to introduce variables and use algebra. We want to prove that $\frac{a\sqrt{a}-b\sqrt{b}}{a\sqrt{b}-b\sqrt{a}} - \frac{a\sqrt{a}+b\sqrt{b}}{a\sqrt{b}+b\sqrt{a}} = 2$. First, let us briefly consider necessary conditions for a and b. Negative numbers for a and b are forbidden from the very beginning because we have to calculate square roots, but also $a = 0$ or $b = 0$ cannot be used because then the denominators would be zero, and, finally, $a = b$ cannot be used because the first denominator would vanish. Altogether we have the conditions $a, b > 0$ and $a \neq b$. First, we will combine the fractions with a common denominator:

$$\frac{(a\sqrt{a} - b\sqrt{b})(a\sqrt{b} + b\sqrt{a}) - (a\sqrt{a} + b\sqrt{b})(a\sqrt{b} - b\sqrt{a})}{(a\sqrt{b} - b\sqrt{a})(a\sqrt{b} + b\sqrt{a})}$$

and then expanding the products appropriately yields

$$\frac{a^2\cancel{\sqrt{ab}} + a^2b - ab^2 - b^2\cancel{\sqrt{ab}} - a^2\cancel{\sqrt{ab}} + a^2b - ab^2 + b^2\sqrt{ab}}{a^2b - b^2a}$$

$$= \frac{2a^2b - 2ab^2}{ab(a - b)} = \frac{2ab(a - b)}{ab(a - b)} = 2$$

Thus, we have shown that the difference of these two fractions is always equal to 2.

Proving an Inequality

In the above problem, one can note that all the fractions of the form $\frac{a\sqrt{a}-b\sqrt{b}}{a\sqrt{b}-b\sqrt{a}}$ seem to be greater than 3. However, is this always true? It is our challenge to prove that this fraction is always greater than 3.

Solution: We have to prove that $\frac{a\sqrt{a}-b\sqrt{b}}{a\sqrt{b}-b\sqrt{a}} > 3$ for all $a, b > 0$ and $a \neq b$. Since we showed above that $\frac{a\sqrt{a}-b\sqrt{b}}{a\sqrt{b}-b\sqrt{a}} - \frac{a\sqrt{a}+b\sqrt{b}}{a\sqrt{b}+b\sqrt{a}} = 2$, our conjecture $\frac{a\sqrt{a}-b\sqrt{b}}{a\sqrt{b}-b\sqrt{a}} > 3$ is equivalent to $\frac{a\sqrt{a}+b\sqrt{b}}{a\sqrt{b}+b\sqrt{a}} > 1$, which, in turn, is equivalent to $a\sqrt{a} + b\sqrt{b} > a\sqrt{b} + b\sqrt{a}$.

Squaring on both sides of this inequality yields $a^3 + b^3 + 2ab\sqrt{ab} > a^2b + b^2a + 2ab\sqrt{ab}$, which can then be written as $(a - b)(a^2 - b^2) > 0$. For $a > b$, both parenthetical expressions are positive, and for $a < b$, both are negative; in both cases the product is positive, and thus, we have proved our conjecture.

Which Is the Larger Number?

Let $a > 1$ be a real number. What is the bigger number $\sqrt{a} - \sqrt{a-1}$ or $\sqrt{a+1} - \sqrt{a}$?

Solution: There is a rather hidden relationship among consecutive square numbers as can be seen here: $1 \xrightarrow{+3} 4 \xrightarrow{+5} 9 \xrightarrow{+7} 16\ldots$. This pattern of differences seems to continue. Might this be the same with square roots? If so, then we might expect that $\sqrt{a+1} - \sqrt{a}$ should increase as the numbers increase. Let us take an example, say, for $a = 9$:

$\sqrt{a}-\sqrt{a-1}=3-\sqrt{8}\approx 0.1716$ and $\sqrt{a+1}-\sqrt{a}=\sqrt{10}-3\approx 0.1623$. It seems to be the reverse! This can be proved in general by using algebra. We have to prove $\sqrt{a}-\sqrt{a-1}>\sqrt{a+1}-\sqrt{a}$, which is equivalent to $2\sqrt{a}>\sqrt{a+1}+\sqrt{a-1}$, and by squaring both sides, and then dividing both sides by 2, we get $2a>a+\sqrt{(a+1)(a-1)}$, and this, in turn, is equivalent to $a>\sqrt{(a+1)(a-1)}$ or $a^2>(a+1)(a-1)=a^2-1$ which proves our conjecture.

To make our understanding of algebra more solid, we should embark on a pursuit of some important algebraic mistakes. Possibly one of the most important rules in mathematics is that one is not allowed to divide by zero. Some people even refer to this as the "eleventh commandment." There are times when division by zero is so well camouflaged that one violates this commandment without knowing it. It is interesting to see what happens when it is breached. Hopefully we will learn from each of these transgressions. What is interesting (or entertaining) is to discover when this rule — dividing by zero — has been violated, thus allowing us to arrive at ridiculous results. Let's now consider a few of these mistakes.

Does 1 = 2? A Mistake Based on Division by Zero

If we square both sides of the equation $a=a$, we get $a^2=a^2$. Then subtracting a^2 from both sides of the equation, we have $a^2-a^2=a^2-a^2$. We will factor the common term a on the left side and factor the difference of two squares on the right side to get $a(a-a)=(a+a)(a-a)$. As $a+a=2a$, this can be rewritten as $a(a-a)=2a(a-a)$. When we now divide both sides of this equation by $a(a-a)$, we get $1=2$. Where did we make the mistake? We have $a-a=0$. Therefore, the above divisor $a(a-a)$ is zero, and we have violated the important rule of not dividing by zero resulting in an absurd statement, namely, $1=2$.

Here is another simple example of this sort of mistake — of dividing by zero — that leads to an absurdity.

Is 1 = 2? A Mistake Based on Division by Zero

This time we will begin with the statement that $a=b$.

We then multiply both sides by b to get:

$$a \cdot b = b \cdot b \quad \text{or} \quad ab = b^2$$

Then subtract a^2 from both sides of the equation so that:

$$ab - a^2 = b^2 - a^2$$

Factoring the common factor on the left and the difference of two squares on the right:

$$a(b - a) = (b + a)(b - a)$$

Dividing both sides by $(b - a)$ give us:

$$a = b + a$$

However, since $a = b$ (which was given), $b = b + b$ or $b = 2b$, which when we divide both sides by b has us resulting with $1 = 2$.

In the following example, division by zero is more camouflaged, thereby making it a bit more difficult to detect.

If $a > b$, Then $a = b$: A Mistake Resulting from Division by Zero

We begin with $a > b$, which can be restated as $a = b + c$, where a, b, and c are positive numbers. We shall now multiply both sides of this equation by $a - b$ to get $a^2 - ab = ab + ac - b^2 - bc$. We then subtract ac from both sides of the equation giving us $a^2 - ab - ac = ab - b^2 - bc$.

Then, factoring both sides of the equations, we get $a(a - b - c) = b(a - b - c)$, and then dividing both sides by $(a - b - c)$ leaves us with $a = b$.

How can it be that $a = b$, when we were told at the beginning that $a > b$? Here we see that, once again, $(a - b - c)$ is equal to 0 because we began with $a = b + c$. We have, thus, violated the rule that forbids dividing by zero.

All Integers are Equal: An Error resulting from Dividing by Zero

Once again, we will use the mistake of dividing by zero — in a somewhat hidden fashion — to show a silly result. We begin by accepting the correct quotient that is as follows: $\frac{x-1}{x-1} = 1$. We now multiply both sides of the equation by $x + 1$ to get $\frac{x^2-1}{x-1} = x + 1$, since $(x + 1)(x - 1) = x^2 - 1$.

Since $(x^2 + x + 1)(x - 1) = x^3 - 1$, we get $\frac{x^3-1}{x-1} = x^2 + x + 1$.

Since $(x^3 + x^2 + x + 1)(x - 1) = x^4 - 1$, we get $\frac{x^4-1}{x-1} = x^3 + x^2 + x + 1$.

And since $(x^{n-1} + x^{n-2} + \cdots + x^3 + x^2 + x + 1)(x - 1) = x^n - 1$, we get $\frac{x^n-1}{x-1} = x^{n-1} + x^{n-2} + \cdots + x^3 + x^2 + x + 1$.

Now suppose we let $x = 1$. The absolute values of the right sides of the above equations equal $1, 2, 3, 4, \ldots, n$.

The left side of the above equations will all be the same, as they are in the form $\frac{1^n-1}{1-1}$, and, therefore, all the right-side numbers must be equal, or $1 = 2 = 3 = 4 = \cdots = n$. Surely, by now you will have realized that the denominators are all $1 - 1 = 0$. This cannot be permitted to exist, since if it did then absurd conclusions would follow, such as that all the numbers $1, 2, 3, \ldots, n$ are equal. We can see here that $\frac{0}{0}$ cannot be a number, or all of these weird results would follow.

The Hidden Nemesis of Dividing by Zero and Not Noticing It Immediately

There are times when the division by zero is well camouflaged. Take, for example, the equation $\frac{3x-30}{11-x} = \frac{x+2}{x-7} - 4$, which allows the right side to be combined as $\frac{3x-30}{11-x} = \frac{x+2-4(x-7)}{x-7}$. This can be then simplified to be $\frac{3x-30}{11-x} = \frac{3x-30}{7-x}$. Since the numerators are equal, the denominators must also be equal, and, therefore, $11 - x = 7 - x$ or $11 = 7$. Quite an absurdity!

It doesn't appear that we divided by zero this time, and yet we ended up with an absurd result.

Had we solved the equation $\frac{3x-30}{11-x} = \frac{3x-30}{7-x}$ in the traditional way, we would find that $x = 10$, which would make the two numerators equal to zero. Still, that doesn't show that we divided by zero.

So, we should consider the following: If $\frac{a}{b} = \frac{a}{c}$, then we multiply both sides by bc to get $ac = ab$. Dividing both sides by a gives us $b = c$, which we expected. However, if $a = 0$, then this would not be valid, since we would have divided by zero.

Let us now return to the equation $\frac{3x-30}{11-x} = \frac{3x-30}{7-x}$, which led us to an absurd result. We found that $x = 10$. With that value of x, the numerators $3x - 30 = 0$, and, therefore, in this case we cannot equate the denominators. Note how slyly division by zero hid from us to deliver a ridiculous result.

More of the Hidden Nemesis: Division by Zero

In a similar vein — but equally well hidden — we can show that $+1 = -1$. We begin with the equation

$$\frac{x+1}{p+q+1} = \frac{x-1}{p+q-1}$$

By subtracting 1 from each side of this equation, we get $\frac{x+1}{p+q+1} - \frac{p+q+1}{p+q+1} = \frac{x-1}{p+q-1} - \frac{p+q-1}{p+q-1}$, which can be simplified to get $\frac{x+1-(p+q+1)}{p+q+1} = \frac{x-1-(p+q-1)}{p+q-1}$, or $\frac{x-p-q}{p+q+1} = \frac{x-p-q}{p+q-1}$.

Since the numerators are equal, the denominators must also be equal so that $p + q + 1 = p + q - 1$ or $+1 = -1$, which is an absurdity! Why did this happen? Might the previous example give a clue?

If you solve the original equation $\frac{x+1}{p+q+1} = \frac{x-1}{p+q-1}$ for x, we find that $x = p + q$.

Therefore, we have the same situation as above, where the numerators of the two equal fractions $(x - p - q)$ were zero.

The initial equation $\frac{x+1}{p+q+1} = \frac{x-1}{p+q-1}$ is not as general as we would at first imagine. It is relevant only for the case where $x = p + q$ and $p + q \neq \pm 1$.

To better understand this result, we can look at a simpler version: From $\frac{a}{b} = \frac{a}{b}$, we cannot simply conclude that $\frac{a+c}{b+c} = \frac{a-c}{b-c}$, since that is only true if:

(1) $a = b$ and $(b + c)(b - c) \neq 0$ or
(2) $c = 0$ and $b \neq 0$.

In other words, we need to make sure that the denominator is not zero.

Finding the Division by Zero Before It Misleads Us

There are many examples of division by zero mistakes that follow a similar pattern. However, division by zero is usually camouflaged and sometimes difficult to find. There are terms that hid the zero so well that it can be easily overlooked, especially when you have no reason to suspect it being there. Let's consider the following example: Suppose a term T_1 is divided by another term $T_2 = \sqrt{4 - 2\sqrt{3}} - \sqrt{3} + 1$; we would not be at all suspicious. However, as you will see in a moment, the term T_2 is of a nature that will violate our now-familiar "eleventh commandment" (thou shall not divide by zero), if we use it as a divisor. In fact, the term T_2 is equal to zero! Follow along the algebra and you will see that it equals zero.

$$\sqrt{4 - 2\sqrt{3}} = \sqrt{3 - 2\sqrt{3} + 1} = \sqrt{(\sqrt{3})^2 - 2 \cdot 1 \cdot \sqrt{3} + 1^2} =$$
$\sqrt{(\sqrt{3} - 1)^2} = \sqrt{3} - 1$, then it follows that

$$T_2 = \sqrt{4 - 2\sqrt{3}} - \sqrt{3} + 1 = 0$$

The zero divisor can be even more hidden as shown in the following:

$$T_3 = \sqrt[3]{\sqrt{5} + 2} + \sqrt[3]{\sqrt{5} - 2} - \sqrt{5}$$

One may now wonder how we can get to show that $T_3 = 0$. Here is a hint that should help you show that $T_3 = 0$.

Note that $\sqrt[3]{\sqrt{5} + 2} = \frac{\sqrt{5}+1}{2}$ and $\sqrt[3]{\sqrt{5} - 2} = \frac{\sqrt{5}-1}{2}$, and $\frac{\sqrt{5}+1}{2} + \frac{\sqrt{5}-1}{2} = \sqrt{5}$. So, now in the value of T_3 when we subtract $\sqrt{5}$, we get 0.

An Absurd Result Stemming from the Well-Known Mistake Based on Division by Zero

Before we begin this example, let's look at a basic principle from algebra. Consider the proportion $\frac{a}{b} = \frac{c}{d}$. From this we can conclude that $\frac{a-c}{b-d} = \frac{c}{d}$, only if $b \neq d$ and $d \neq 0$. To show that this rule actually is true, we begin by recognizing that from the original proportion $ad = bc$ (by cross multiplication). The cross multiplication for the second proportion (above) gives us the following: $(a - c)d = (b - d)c$ or $ad - cd = bc - dc$, which, when adding cd to both sides, gives us $ad = bc$. This is the same as we had gotten for the first proportion. Now having established the rule above, we shall apply it to the following situation: Given x, y, z and the proportion $\frac{3y-4z}{3y-8z} = \frac{3x-z}{3x-5z}$, we shall now apply the rule we established above to get $\frac{3y-4z-(3x-z)}{3y-8z-(3x-5z)} = \frac{3x-z}{3x-5z}$, and then $\frac{3y-4z-3x+z}{3y-8z-3x+5z} = \frac{3x-z}{3x-5z}$, which simplifies to $\frac{3y-3z-3x}{3y-3z-3x} = 1 = \frac{3x-z}{3x-5z}$. It then follows that $3x - 5z = 3x - z$, which gives us $5 = 1$.

There must be something wrong, since we ended up with an absurd result.

The mistake here is a bit more difficult to find. The equation $\frac{3y-4z}{3y-8z} = \frac{3x-z}{3x-5z}$ is satisfied when $x - y + z = 0$. Now substituting $x = y - z$, we get $\frac{3x-z}{3x-5z} = \frac{3(y-z)-z}{3(y-z)-5z} = \frac{3y-4z}{3y-8z}$.

Using a similar substitution, we have $3y - 3z - 3x = 3y - 3z - 3(y-z) = 3y - 3z - 3y - +3z = 0$ and $3y - 3z - 3x = 3y - 8z + 5z - 3x = 3y - 8z - (3x - 5z) = 0$. Both are equal to zero. Thus, the denominator of the fraction $\frac{3y-4z-(3x-z)}{3y-8z-(3x-5z)}$ is equal to zero. This is an example of where division by zero is well camouflaged and exemplifies the kind of subtle mistakes that mathematicians have to be cautious about throughout their investigations.

Absurd Results Based on a Mistake of Interpretation

We are asked to solve the system of equations:

(1) $a + b = 1$
(2) $a + b = 2$.

Our initial reaction is to note that since the left sides of these equations are equal, then so must the right sides be equal. Thus, we find that $1 = 2$. "Proved!" Or is it?

We could also have embarked on this system of equations by subtracting the two equations — knowing that the difference of equals is also equal. When we subtract the first equation from the second equation, on the left side we would get 0 and on the right side we would get 1. Thus, $0 = 1$. Once again, an absurd result.

To take this one step further, with this set of equations we could also show that $1 = -1$. When we subtracted the first equation from the second equation, we got $0 = 1$. If we, now, subtract the second equation from the first equation, we get $0 = -1$. We can take this absurdity even one foolish step further.

Since $0 = 1$ and $0 = -1$, one could then conclude that $1 = -1$, as both are equal to 0.

Our series of absurd conclusions above result from the fact that these two equations have no common solution. Were we to graph them, they would appear as two parallel lines — thus having no intersection or point in common. The mistake here was to embark on the two equations, seeking a solution and not recognizing immediately that there cannot be a solution when the two lines representing these equations are parallel and, therefore, have no common point of intersection.

Some mistakes in algebra can be seen better graphically, as we can see in Figure 4.17. Consider the two equations $5x + y = 15$ and $x = 4 - \frac{y}{5}$. If we substitute the value of x from the second equation into the first equation, we will get $5\left(4 - \frac{y}{5}\right) + y = 15$. This simplifies to $20 - y + y = 15$ or $20 = 15$. Now, there must be something clearly wrong here. Where was the mistake? If we multiply both sides of the second of the two given equations by 5, we get $5x + y = 20$. Were we to graph these two equations, we would find them to be parallel and would, therefore, have no point of intersection, or to put it another way, no common solution (see Figure 4.17). Therefore, it makes no sense to try to solve these two equations simultaneously as we did above — thus, clearly, leading to an absurd result!

This time the parallelism of the two equations was not as obvious as in the first case above. Yet, to avoid such absurd results, we have to be cautious not to make some of the mistakes of interpretation shown here.

With this reasoning we can also prove that $5 = 16$. To do this, we begin with the equation

$$(x + 1)^2 - (x + 2)(x + 3) = (x + 4)(x + 5) - (x + 6)^2$$

Then doing the indicated multiplications, we get

$$x^2 + 2x + 1 - (x^2 + 5x + 6) = x^2 + 9x + 20 - (x^2 + 12x + 36)$$

By combining like terms, we then get $-3x - 5 = -3x - 16$.

Adding $3x$ to both sides gives us $-5 = -16$.

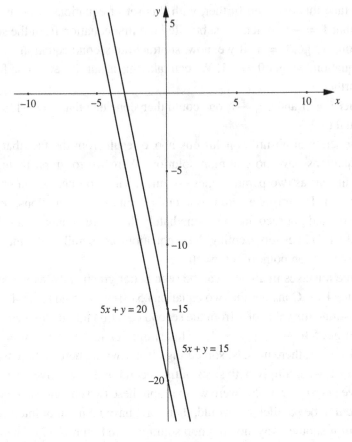

Figure 4.17

Lastly, multiplying by (-1), we get the absurd result that $5 = 16$. From a given equation the mathematical mistake is less obvious, but as you approach the ridiculous result, you will note the similarity to the previous oversights. You then ask yourself, where is the mistake? We multiplied and added correctly. The mistake lies in the original equation, which only becomes clear at the end of this discussion. In other words, assuming the existence of a solution (but there is none) of this last equation, we find the contradiction $5 = 16$. From a contradiction, we conclude that one of our assumptions or some reasoning must be wrong. But here all steps are correct except our assumption that the equation has a solution. Therefore, we can conclude that our original equation has no solution.

Simultaneous Equations Leading to a Strange Result through an Algebraic Mistake

We begin with the following two equations to be solved as a pair of simultaneous equations:

$$\frac{x}{y} + \frac{y}{x} = 2 \tag{1}$$

$$x - y = 4 \tag{2}$$

Then by multiplying equation (1) by xy, we arrive at $x^2 + y^2 = 2xy$, which can then be reworked to give us the following: $x^2 - 2xy + y^2 = 0$.

This can be written as $(x - y)^2 = 0$, whereupon $x - y = 0$, or put another way, $x = y$.

If we substitute this value of y in equation (2), we end up with the ridiculous result $x - x = 0 = 4$, which then results in $0 = 4$. So where is the mistake?

From the given we would assume that x and y are not zero. Actually, this system of equations has no solution. If we look at this graphically, we will note that we actually have two parallel lines. With no intersection of the two lines there can be no common solution as can be seen graphically in Figure 4.18.

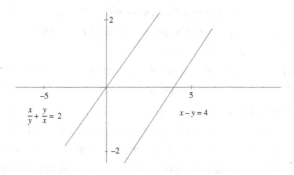

Figure 4.18

A Mistake Based on Faulty Square Root Extraction Leading to an Absurd Conclusion

In this case just follow the steps shown in the following:

$$2 = 2$$

$$3 - 1 = 6 - 4$$

$$1 - 3 = 4 - 6$$

$$1 - 3 + \frac{9}{4} = 4 - 6 + \frac{9}{4}$$

$$1 - 2 \cdot \frac{3}{2} + \frac{9}{4} = 4 - 4 \cdot \frac{3}{2} + \frac{9}{4}$$

$$\left(1 - \frac{3}{2}\right)^2 = \left(2 - \frac{3}{2}\right)^2$$

$$1 - \frac{3}{2} = 2 - \frac{3}{2}$$

$$1 = 2$$

So where is the error?

It is hidden in the step where we took the square root of both sides of the following $\left(1 - \frac{3}{2}\right)^2 = \left(2 - \frac{3}{2}\right)^2$, and arrived at this expression $1 - \frac{3}{2} = 2 - \frac{3}{2}$, which ignored the negatives that need to be considered. We should have gotten the absolute values as follows: $\left|1 - \frac{3}{2}\right| = \left|2 - \frac{3}{2}\right|$, which would have led us to $\left|-\frac{1}{2}\right| = \left|\frac{1}{2}\right|$, which results in something that is certainly reasonable (and correct) $\frac{1}{2} = \frac{1}{2}$. Ignoring the proper square root extraction can lead to countless mistaken and silly results.

For example, suppose we begin with $-20 = -20$, and write it as $16 - 36 = 25 - 45$. If we add $\frac{81}{4}$ to both sides, we get $16 - 36 + \frac{81}{4} = 25 - 45 + \frac{81}{4}$, which is equivalent to $\left(4 - \frac{9}{2}\right)^2 = \left(5 - \frac{9}{2}\right)^2$. Now taking the square root of both sides (although incorrectly, as stated above) we get the following: $4 - \frac{9}{2} = 5 - \frac{9}{2}$, which then results in the ridiculous result $4 = 5$.

However, had we done the work correctly by considering that the square root should result in an absolute-value statement, we would have gotten the following: $\left|4 - \frac{9}{2}\right| = \left|5 - \frac{9}{2}\right|$, which leads to a sensible result $\frac{1}{2} = \frac{1}{2}$.

A Subtle Mistake in Solving an Equation Will Cause an Error

We are asked to solve the equation $3x - \sqrt{2x - 4} = 4x - 6$ (where $x \in$ Real Numbers, $x \geq 2$). The usual way to solve this equation is to isolate the radical term and then square both sides as follows:

$$\sqrt{2x - 4} = -(x - 6)$$
$$2x - 4 = (-(x - 6))^2 = (x - 6)^2$$

Simplifying this we get $x^2 - 14x + 40 = 0$. This then produces two roots $x_1 = 10$ and $x_2 = 4$. By substituting these two values into the original equation, we find that only x_2 is a solution and not x_1. Why is x_1 not a valid solution? The mistake we made was to consider taking the square root of both sides, which is not an equivalent change. In other words, from $T_1 = T_2$ it follows that $T_1^2 = T_2^2$. However, the reverse is not true — as we have just experienced. In Figure 4.19 we graph each side of the original equation and can see where the two equations intersect.

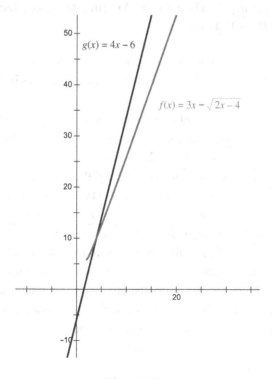

Figure 4.19

To "Prove" That $0 = 100$: Stemming from a Mistake with Square-Root Extraction

We begin by letting $y = 100$ and $z = 0$. Suppose we have $x = \frac{y+z}{2}$. Then $2x = y + z$.

By multiplying both sides by $y - z$, we get $2x(y - z) = (y+z)(y-z)$. Now multiply as indicated: $2xy - 2xz = y^2 - z^2$. Then rearrange terms: $z^2 - 2xz = y^2 - 2xy$. Then adding x^2 to both sides of the equation: $z^2 - 2xz + x^2 = y^2 - 2xy + x^2$. Factor each of the sides of the equation: $(z - x)^2 = (y - x)^2$. Taking the square root of each side gives us: $z - x = y - x$. This leaves us with $z = y$, or by substituting the original values for y and z, we have $0 = 100$. Once again, the mistake lies in concluding that $a^2 = b^2$ implies that $a = b$. But the conclusion can only be $a = \pm b$. So that in the next-to-last step we could have gotten $z - x = -(y - x)$, which then gets us back to where we began with $x = \frac{y+z}{2}$.

To Show That If $a \neq b$, Then $a = b$: Another Mistake Stemming from Square-Root Extraction

There are times when our mistake could be of the same nature as that above, but well hidden, and therefore, easily overlooked as in the following example.

We begin with $a \neq b$ and assume (without loss of generality) $a < b$. Further, let be $c = \frac{a+b}{2}$. That means that $a + b = 2c$. By multiplying both sides by $a - b$, we get $a^2 - b^2 = 2ac - 2bc$.

Then adding $b^2 - 2ac + c^2$ to both sides of the previous equation, we get $a^2 - 2ac + c^2 = b^2 - 2bc + c^2$. Each of the two sides of this equation is a perfect square, and, thus, the equation can be written as $(a - c)^2 = (b - c)^2$. Taking the square root of both sides $\sqrt{(a - c)^2} = \sqrt{(b - c)^2}$ gives us $a - c = b - c$ or $a = b$. Recall that we began by stating that $a \neq b$. There must be a mistake somewhere in our work. We seem to have done every step of this algebraic process correctly. However, in the last step, where we took the square root of both sides, we neglected to consider the negative values. Had we taken the result of extracting the square root of both sides of $\sqrt{(a-c)^2} = \sqrt{(b-c)^2}$ to get $a - c = -(b-c)$, we would have gotten $a - c = -b + c$, which is our original equation, $a + b = 2c$.

Caution Must be Taken in Solving Equations or Else a Mistake May be Encountered

Another error arising from a surprise mistake is one very subtly hidden in the procedure of solving the equation $1 + \sqrt{x+2} = 1 - \sqrt{12-x}$.

We begin our solution to this equation by adding -1 to both sides and then squaring the two sides. This yields $x + 2 = 12 - x$, which results in $x = 5$. If we substitute this value of x into the original equation, we get $1 + \sqrt{5+2} = 1 - \sqrt{12-5}$, and then adding -1 to both sides and then squaring both sides of $\sqrt{5+2} = -\sqrt{12-5}$, we get $7 = 7$. This would have us think that the value of x is the correct value. It is not! If we substitute 5 in place of x in the original equation, we will have $1 + \sqrt{7} = 1 - \sqrt{7}$, which is not correct. There is no answer to this equation. This can be seen from the very beginning, since the initial equation is equivalent with $\sqrt{x+2} = -\sqrt{12-x}$. The only way a square root is a negative square root is that the radicand is 0. But this would imply on the left side $x = -2$ and on the right side $x = 12$.

Where, then, has the mistake been made? When taking the square root, we must take the positive *and* the negative into account. We violated that rule in this process!

It is good to remember that squaring both sides of an equation is not an equivalence transformation. It yields a new equation with possibly more solutions than the given equation. Therefore, not every solution of the "squared equation" is a solution of the original equation. This is an important rule very often not explicitly expressed. This is when mistakes appear.

Consider the equation $x + 5 - \sqrt{x+5} = 6$. This can be written as $x - 1 = \sqrt{x+5}$. Squaring both sides and solving, we get $x^2 - 2x + 1 = x + 5$, or simplified as $x^2 - 3x - 4 = 0$. Then $x = 4$ or $x = -1$. However, whereas $x = 4$ is a solution, $x = -1$ is not a solution. This is a typical mistake made in algebra classes. Again, the square root process did not take the negative into account.

This absurdity can be taken further, for if we want to prove that $5 = 1$, we then subtract 3 from both sides to get $2 = -2$, and then squaring both sides, we get $4 = 4$. Therefore, 5 must have equaled 1!

A Mistake with Powers Can Lead Awry

Imagine someone solving the equation $\left(\frac{2}{3}\right)^x = \left(\frac{3}{2}\right)^3$ as follows: Begin with the given $\left(\frac{2}{3}\right)^x = \left(\frac{3}{2}\right)^3$ and then apply the powers to the fractions to get $\frac{2^x}{3^x} = \frac{3^3}{2^3}$. Multiplying by the common denominator, $2^3 \cdot 3^x$ leaves us with $2^3 \cdot 2^x = 3^3 \cdot 3^x$.

Following the rules of exponents, we have $2^{3+x} = 3^{3+x}$.

If two equal powers have equal exponents, we might conclude that the bases must be equal as well, therefore, $2 = 3$. Something is not correct here. Where might the mistake be? The last step is wrong! The correct solution is $x = -3$, which makes each of the two sides equal to 1.

You may want to see what this looks like graphically; we provide the two functions $f(x) = 2^{3+x}$ and $g(x) = 3^{3+x}$ and show the graph in Figure 4.20.

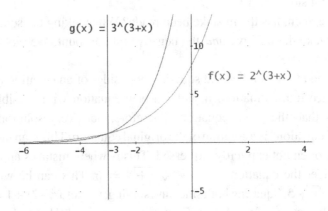

Figure 4.20

A Subtle Oversight — or Mistake — with the Binomial Theorem

We know that $(a + b)^2 = a^2 + 2ab + b^2$. This formula is an application of the binomial theorem, which provides a formula for obtaining the value of a binomial raised to *any* positive integer power n. The formula is

$$(a + b)^n = a^n + na^{n-1}b + \frac{n(n - 1)}{2}a^{n-2}b^2 +$$

$$\cdots + \frac{n(n - 1)}{2}a^2b^{n-2} + nab^{n-1} + b^n$$

If $n = 2$, we get $(a + b)^2 = a^2 + 2ab + b^2$. If $n = 1$, we get $(a + b)^1 = a^1 + b^1 = a + b$. Using the formula of the binomial theorem, we can also be led by mistake to a weird conclusion. When $n = 0$, we get $1 = 1 + 0 + 0 + 0 + \cdots + 0 + 0 + 0 + 1$, or $1 = 2$. (Because when $n = 0$, then $(a + b)^0 = 1$, $a^0 = 1$, $b^0 = 1$.) That we reached an absurd result would have us believe that a mistake was made, and indeed one was. But where is the error? This seems to be alright: $(a + b)^n = (a + b)^0 = 1$. Now look at the right side more precisely, for therein lies the reason for our mistake. On the right side of the equation, there is for $n = 0$ only *one* term (not *two or more* terms), which we produced by substituting into the binomial formula. That term is $1a^{0-0}b^0 = 1 \cdot a^0b^0 = 1 \cdot 1 \cdot 1 = 1$. We can also see that on the Pascal triangle, which can be used to determine the coefficients of the terms of the binomial expansion as follows and shown in Figure 4.21.

$$(a + b)^0 = 1$$

$$(a + b)^1 = 1a + 1b$$

$$(a + b)^2 = 1a^2 + 2ab + 1b^2$$

$$(a + b)^3 = 1a^3 + 3a^2b + 3ab^2 + 1b^3$$

$$\cdots$$

$$(a + b)^n = 1a^n + \frac{n}{1}a^{n-1}b + \frac{n(n - 1)}{2}a^{n-2}b^2 + \cdots + \frac{n(n - 1)}{2}a^2b^{n-2} + \frac{n}{1}ab^{n-1} + 1b^n$$

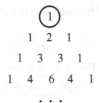

$$1 \quad 2 \quad 1$$
$$1 \quad 3 \quad 3 \quad 1$$
$$1 \quad 4 \quad 6 \quad 4 \quad 1$$
$$\cdots$$

Figure 4.21

But be careful: If $a = b = 0$, you could have a problem! What is the value of 0^0? This is an expression that we leave undefined, and many calculators do not react to this input, yet some deliver the value 1, since "anything" taken to the zero power seems to be 1.

As we said at the outset, the binomial theorem only holds for *positive* values of n. Note the mistake that using zero has caused. That is why we specify the value of n to avoid such ridiculous results!

To Show That If p Is Positive, It Is Actually Negative: A Mistake with Inequalities

We begin by assuming that both p and q are positive, and we are going to show that p is negative. Clearly, the inequality $2q - 1 < 2q$ is a true statement. Suppose we multiply both sides by $-p$ to get $-2pq + p < -2pq$. If we then add $2pq$ to both sides of this inequality, we end up with $p < 0$, which is to say p is negative. How can this be when we started off with a positive p? Where was the mistake?

We violated a rule for inequalities as follows:

When multiplying (or dividing) both sides of an inequality by a negative number, the inequality symbol must be reversed.

Look at a simple example: $2 < 3$. But, when we multiply both sides of the inequality by -1, we have the following: $2 \cdot (-1) = -2$, which is greater than $3 \cdot (-1) = -3$, or simply written $-2 > -3$.

We can see this mistake played out in a less obvious way in the following example.

To Show That Any Positive Number Is Greater Than Itself

We shall begin with the two positive numbers p and q, where $p > q$. We will now multiply both sides of this inequality by q to get $pq > q^2$. Now subtracting p^2 from both sides of the inequality gives us $pq - p^2 > q^2 - p^2$. By factoring both sides, we get $p(q - p) > (q + p)(q - p)$. Dividing both sides by $(q - p)$ leaves us with $p > q + p$, which says that p is greater than itself. That's absurd! So where was the mistake made? Since $p > q$, it must follow that $(q - p)$ is negative. We made the mistake of not reversing the inequality symbol when dividing both sides of the inequality by the negative term $q - p$.

The structure of this example follows the "proof" of "$1 = 2$: A mistake based on division by zero" — instead of division by zero in an equation we had a division by a negative term for an inequality. This absurd result can be taken a step further. With the given $p > q$ and our newly found result $p > q + p$, we can add the two inequalities to get $2p > 2q + p$. Now subtracting p from both sides, we get $p > 2q$. So if having $p > q$, and $p > 2q$, we can conclude that $p > 2q$, similar reasoning would allow us to then conclude that $p > 4q$. This can continue in the same way to further absurdities.

Mistaken Application of Complex Numbers to Prove That $-1 = +1$

We begin with the product of two imaginary numbers $\sqrt{-1}$ and apply the rules we know about real numbers: $\sqrt{-1} \cdot \sqrt{-1} = \sqrt{(-1) \cdot (-1)} = \sqrt{+1} = 1$. This time we will evaluate the product as follows: $\sqrt{-1} \cdot \sqrt{-1} = (\sqrt{-1})^2 = -1$. Therefore, since the given product provides us with two values, we would have to conclude that $-1 = +1$. Something must be wrong, since clearly -1 is not equal to $+1$. This mistake is one that depends on a definition in mathematics — one that we may even think arose to avoid this dilemma. That is, that the product $\sqrt{a} \cdot \sqrt{b} = \sqrt{a \cdot b}$ does not hold true when a and b are negative. Therefore, in a sense $\sqrt{-1} \cdot \sqrt{-1} = \sqrt{(-1) \cdot (-1)}$ is wrong! However, it is true that $\sqrt{(-1) \cdot (-1)} = \sqrt{1}$. Another way to avoid (or explain) the above dilemma is the following: In the *real numbers*, taking a square root is a process with a *unique* result; solving an equation may have more solutions, for example, $\sqrt{4} = 2$ but the equation $x^2 = 4$ has the two solutions ± 2. In the world of *complex numbers*, this is not true any longer, square roots always have two possible values, cube roots always have three possible values, and so on. In terms of complex numbers solving the equation $x^2 = 4$ is the same as taking a square root, which is a big difference between real and complex numbers! So, there is no unique value of $\sqrt{-1}$, it is not only i,[1] rather we have $\sqrt{-1} = \pm i$. And seeing it this way, the above equation does not produce a contradiction, it merely says $(\pm i) \cdot (\pm i) = 1$ which is correct, because the left side is nothing other than $\pm i^2 = \pm(-1)$. Analogously, a similar dilemma arises when we take for granted the rule for real numbers in the quotient operation as $\frac{\sqrt{a}}{\sqrt{b}} = \sqrt{\frac{a}{b}}$ and carelessly extend it to negative numbers.

The following is clearly true since both sides of the equation are equal to $\sqrt{-1}$. Now observe what happens when we accept the above generalization: Starting with $\sqrt{\frac{1}{-1}} = \sqrt{\frac{-1}{1}}$ would then lead to $\frac{\sqrt{1}}{\sqrt{-1}} = \frac{\sqrt{-1}}{\sqrt{1}}$.

Now clearing fractions (perhaps by either multiplying by the common denominator or simply cross-multiplying), we get $(\sqrt{1})^2 = (\sqrt{-1})^2$. This essentially tells us that $1 = -1$. Again, the definition was abused leading to a mistaken result. To debunk this "proof," one need not know much about complex numbers, just a familiarity with the characteristics of the familiar operations. We note that there are times when our time-honored operations take on other characteristics.

A Subtle Mistake Leads to an Absurdity

Follow along as we solve this equation for x (a real number): $\frac{6}{x-3} - \frac{9}{x-2} = \frac{1}{x-4} - \frac{4}{x-1}$.

---

[1] Where i is the famous imaginary number with $i^2 = -1$.

We begin by multiplying the fractions on each side by their respective lowest common denominator:

$$\frac{6(x-2)}{(x-2)(x-3)} - \frac{9(x-3)}{(x-2)(x-3)} = \frac{x-1}{(x-1)(x-4)} - \frac{4(x-4)}{(x-1)(x-4)}.$$

We then clear parentheses and add the fractions on each side of the equation: $\frac{6x-12-9x+27}{x^2-3x-2x+6} = \frac{x-1-4x+16}{x^2-4x-x+4}$. By combining like terms, we get $\frac{-3x+15}{x^2-5x+6} = \frac{-3x+15}{x^2-5x+4}$. Now divide both sides by $(-3x + 15)$ to get $\frac{1}{x^2-5x+6} = \frac{1}{x^2-5x+4}$. We then equate denominators, since the numerators and the fractions are equal: $x^2 - 5x + 6 = x^2 - 5x + 4$. By subtracting $x^2 - 5x$ from both sides of the equation, we end up with $6 = 4$.

With this absurd result, you would think that the original equation has no solution. This is wrong! The solution of this equation is $x = 5$, as you can see from the following, where we show that when $x = 5$, each side of the original equation has the same value, namely, 0:

$$\frac{6}{5-3} - \frac{9}{5-2} = \frac{6}{2} - \frac{9}{3} = 3 - 3 = 0 \quad \text{and}$$

$$\frac{1}{5-4} - \frac{4}{5-1} = \frac{1}{1} - \frac{4}{4} = 1 - 1 = 0$$

We note that x cannot take on the values of 1, 2, 3, and 4, since that would produce a zero denominator in one of the fractions of the original equation. So then, where might the error lie? When we divided by $-3x + 15$, we had to eliminate the possibility that $-3x + 15 = 0$. However, this case is the one that provides us with the correct answer since $3x = 15$, and, therefore, $x = 5$. Therefore, we once again — surprisingly — divided by zero. Our old nemesis!

A Confusing Equation: Ripe for a Mistake (or an Omission)

We seek to solve the following equation for the real number x: $3 - \frac{2}{1+x} = \frac{3x+1}{2-x}$.

Add the terms on the left side of the equation to get $\frac{3(1+x)}{1+x} - \frac{2}{1+x} = \frac{3x+1}{1+x}$. This gives us $\frac{3x+1}{1+x} = \frac{3x+1}{2-x}$. Since the numerators are equal, the denominators must be equal as well so that $1 + x = 2 - x$. Solving for x, we get $2x = 1$ and $x = \frac{1}{2}$. This would appear to be a solution to the equation. Let's check to see if this checks out properly. Substituting the $\frac{1}{2}$ for x we get the following: The left side of the equation: $3 - \frac{2}{1+\frac{1}{2}} = 3 - \frac{2}{\frac{3}{2}} = 3 - \frac{4}{3} = \frac{5}{3}$. The right side of the equation: $\frac{3 \cdot \frac{1}{2}+1}{2-\frac{1}{2}} = \frac{\frac{5}{2}}{\frac{3}{2}} = \frac{5}{3}$. All appears to be fine. Unfortunately, that is not the only solution to this equation. Let's begin another method for solving this equation: $3 - \frac{2}{1+x} = \frac{3x+1}{2-x}$. Multiply both sides of the equation by $(1+x)(2-x)$ to get $3 \cdot (1+x)(2-x) - 2 \cdot (2-x) = (3x+1)(1+x)$. Then clearing parentheses, we get $-3x^2 + 3x + 6 - 4 + 2x = 3x^2 + 3x + x + 1$. Then simplifying: $-3x^2 + 5x + 2 = 3x^2 + 4x + 1$. We will now add $3x^2 - 5x - 2$ to both sides of the equation to get $0 = 6x^2 - x - 1$. By dividing both sides of the equation by 6, we get $x^2 - \frac{1}{6}x - \frac{1}{6} = 0$.

Using the well-known quadratic formula to solve this quadratic equation, we get the two values for x: $x_{1,2} = \frac{1}{12} \pm \sqrt{\frac{1}{12^2} + \frac{1}{6}} = \frac{1}{12} \pm \sqrt{\frac{1}{144} + \frac{24}{144}} = \frac{1}{12} \pm \frac{5}{12}$. Or, written separately: $x_1 = \frac{1}{12} + \frac{5}{12} = \frac{1}{2}$ and $x_2 = \frac{1}{12} - \frac{5}{12} = -\frac{1}{3}$. Then $x_2 = -\frac{1}{3}$ is then a second solution to this equation (along with the previously found $x_1 = \frac{1}{2}$).

We ought to check the second solution to see if it is, in fact, a correct solution. Substituting the second solution, $x_2 = -\frac{1}{3}$, on the left side of the original equation: $3 - \frac{2}{1-\frac{1}{3}} = 3 - \frac{2}{\frac{2}{3}} = 3 - 3 = 0$. Then substituting the value $x_2 = -\frac{1}{3}$ on the right side of the equation, we have $\frac{3(-\frac{1}{3})+1}{2+\frac{1}{3}} = \frac{0}{\frac{7}{3}} = 0$. So clearly the second solution is also correct. Our mistake here was to solve an equation and to be perhaps satisfied with only one of two possible correct solutions.

An Equation That Leads to a Mistaken Solution

Suppose we search for the solution (x, y) of the equation $x^2 + 9y^2 = 0$, where x and y are the real numbers. Follow the following solution: $x^2 + 9y^2 = 0$. Subtract x^2 from both sides of the equation: $9y^2 = -x^2$. Divide both sides equation by 9: $y^2 = -\frac{x^2}{9}$. Take the square root of both sides of the equation: $y = \pm\frac{x}{3}$. This is a mistake and not the correct solution.

Since $x^2 \geq 0$ and $9y^2 \geq 0$, we know that $x^2 + 9y^2 \geq 0$. However, we are given that $x^2 + 9y^2 = 0$. Therefore, it must follow that $x^2 = 0$ and $9y^2 = 0$, which implies that $x = 0$, as well as $y = 0$, and not $y = \pm\frac{x}{3}$.

(The original equation has its only solution $x = 0 = y$ in the real numbers yet in the complex numbers, the following is a solution: $y = \frac{i}{3}x$ or $y = -\frac{i}{3}x$.)

Caution Must Be Taken When Solving Inequalities — Otherwise We Will Be Making a Mistake

We begin with the real numbers a and b. For which values of a and b is the following inequality satisfied? $\frac{a}{b} + \frac{b}{a} > 2$. We realize that the values of a and b cannot be zero, or else the division indicated would be invalid. We begin by multiplying both sides by ab to get: $a^2 + b^2 > 2ab$. Then add to both sides the following: $-ab - b^2$ to get: $a^2 - ab > ab - b^2$.

Factoring the common factor on both sides of the equation gives the following: $a(a - b) > b(a - b)$. Our last step is to divide both sides by $(a - b)$, which results in $a > b$. The solution to this inequality appears to be $a > b$. Is this answer correct? Clearly, with the division by $(a - b)$, we realized that $a \neq b$. If $a = b$ then following would be true: $\frac{a}{a} + \frac{a}{a} = 1 + 1 = 2$, which contradicts the given inequality.

So now let's look at an alternate solution to this inequality: $\frac{a}{b} + \frac{b}{a} > 2$. We begin by multiplying both sides by ab. If, in the first case, $ab > 0$ (that means a and b have the same sign; and notice that $a \neq b$) then we get $a^2 + b^2 = 2ab$ which is equivalent to $(a - b)^2 > 0$, which is true for every such a, and b. In the second case, $ab < 0$ (that means a and b have different signs), we would end up with $(a - b)^2 < 0$ which is never true. The correct answer is the following: (a) $a \neq b$ and $a, b > 0$, and (b) $a \neq b$ and $a, b < 0$. In other words, when a and b have different signs, the inequality is not satisfied. This can be seen graphically in Figure 4.22.

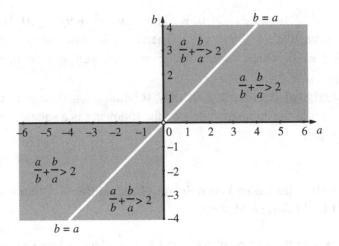

Figure 4.22

More Mistakes That Lead to Correct Results

As we have seen from the previous example, mistakes don't always lead to an absurd result. We could also have mistakes that lead to a correct answer. These are not to be condoned but just provide us with some amusement.

We begin with the equation $x - 2 = 3$, which is the same as $x = 5$. Now we will make a deliberate mistake and add 12 to *only* the left side of the original equation to get $x + 10 = 3$. Then we will multiply both sides of the equation by $x - 5$ to get $(x + 10)(x - 5) = 3(x - 5)$. We now subtract $3(x - 5)$ from both sides of the equation, which gives us $x^2 + 5x - 50 - (3x - 15) = 0$ or in its simplified form, $x^2 + 2x - 35 = 0$. By factoring, we get $(x + 7)(x - 5) = 0$. Dividing both sides by $x + 7$ gives us $x - 5 = 0$ or $x = 5$, which is what we had as our initial value of x. So, despite our earlier mistake of adding 12 to only one side of the equation, we still got the right result.

Had we not added 12 to only one side of the equation, but to both sides, as we should have, we would have subtracted $15(x - 5)$ instead of $3(x - 5)$. This would have given us $(x - 5)^2 = 0$, implying that $x = 5$. The "wrong" solution $x = -7$ will have disappeared through the division by $x + 7$.

Another comical mistake that leads to a correct answer is to incorrectly *add* the two binomials instead of the indicated multiplication.

We begin with the equation that we are asked to solve for x to get $(5 - 3x)(7 - 2x) = (11 - 6x)(3 - x)$. Now *adding* instead of *multiplying*, as is indicated in the given equation, we get $(5 - 3x) + (7 - 2x) = (11 - 6x) + (3 - x)$. This can (correctly) be converted to $12 - 5x = 14 - 7x$. This yields $2x = 2$ or $x = 1$, which, surprisingly, is correct!

Compare this to solving the equation $(5 - 3x)(7 - 2x) = (11 - 6x)(3 - x)$ correctly, which leads to $6x^2 - 31x + 35 = 6x^2 - 29x + 33$, which has the sole solution $x = 1$.

Here is a rather silly series of two mistakes that leads to the correct answer: $\sqrt{\frac{2.8}{70}} = \sqrt{0.04} = 0.2$. In other words, the second error corrects the first error.

A Correct Start Followed by a Series of Silly Mistakes Leads to Correct Solution

We are given the following equation and asked to solve for x: $\frac{x-7}{x+7} + \frac{x+10}{x+3} = 2$. But multiplying both sides of the equation by $(x+7)(x+3)$, as we would normally begin in solving this equation, we get the following.

$$(x - 7) \cdot (x + 3) + (x + 10) \cdot (x + 7) = 2(x + 7) \cdot (x + 3)$$

$$x - 7 + x + 7 = 2$$

$$2x = 2$$

$$x = 1$$

This is then followed by the mistaken cancellations shown in the following, which is then followed by some further sloppiness that brings us to a correct answer. This mistake may appear comical but has been already committed by some.

Mistake for Which We Can Blame the Calculator

Some mistakes in mathematics may not be our fault. Rather, they may be the fault of the calculator on which we seem to have unquestioned reliance. Suppose we take an algebraic fraction such as $\frac{1}{\sqrt{a+b}-\sqrt{a}}$ and correctly find its equivalent by first multiplying by 1 in the form of $\frac{\sqrt{a+b}+\sqrt{a}}{\sqrt{a+b}+\sqrt{a}}$ and then doing the algebra as shown here:

$$\frac{1}{\sqrt{a+b}-\sqrt{a}} = \frac{1}{\sqrt{a+b}-\sqrt{a}} \cdot \frac{\sqrt{a+b}+\sqrt{a}}{\sqrt{a+b}+\sqrt{a}} = \frac{\sqrt{a+b}+\sqrt{a}}{(\sqrt{a+b})^2 - (\sqrt{a})^2}$$

$$= \frac{\sqrt{a+b}+\sqrt{a}}{(a+b) - a} = \frac{\sqrt{a+b}+\sqrt{a}}{b}$$

Let us now compare the way the calculator evaluates the two equal algebraic expressions:

$$\frac{1}{\sqrt{a+b}-\sqrt{a}} \quad \text{and} \quad \frac{\sqrt{a+b}+\sqrt{a}}{b}$$

Assigned values	Calculator result to n places	$\frac{1}{\sqrt{a+b}-\sqrt{a}}$	$\frac{\sqrt{a+b}+\sqrt{a}}{b}$
$a = 1,000$	8	63,291.**139**	63,245.**569**
$b = 0.001$	20	63,245.569014751**992618**	63,245.569014751**934636**
$a = 100$	8	2000.**4001**	2000.**05**
$b = 0.01$	20	2000.0499987500624**968**	2000.0499987500624**960**

Note the differences — or we should say, note the mistakes — that the calculator has made. These are not exactly mathematical mistakes. They are rounding-off mistakes that result from the fact that the first denominator is a difference of approximately equal numbers. And such differences in a denominator are, numerically, seen to be very dangerous! The more accurate result we get from the second fraction.

Mistaken Relationships

It is an unfortunate mistake when one misunderstands what a proportion is. The common mistake is one that can be seen when a woman is asked her age and reasons as follows: I was 20 years old when I got married to my husband, who at that time, was 30 years old. Since he is 60 years old today, which is twice 30, I suspect that I am 40 years old, which is twice 20.

In other words, the woman figured as follows: $\frac{x}{20} = \frac{60}{30}, x = 20 \times 2 = 40$.

Unfortunately, this is a mistake. This cannot be handled as a proportion, since it merely requires a constant difference. In this case, there is a difference of ten years; therefore, her correct age is $60 - 10 = 50$.

Another Absurdity

In a proportion, if the first term is greater than the second term, then the third term must be greater than the fourth term. Therefore, if $ad = bc$, then $\frac{a}{b} = \frac{c}{d}$. Suppose $a > b$, then $c > d$.

Now if we let $a = d = 1$, and $b = c = -1$, we have satisfied the equation $ad = bc$, where $a > b$. It then should follow that $c > d$, which in this case would indicate that $-1 > 1$. This is clearly a mistake, but where is the error?

The mistake here is that we said earlier that $\frac{a}{b} = \frac{c}{d}$, and suppose $a > b$, then $c > d$. This only holds true for positive numbers but not in general. For example, if $\frac{a}{b} = \frac{c}{d}$, and $a > b$, we could have $\frac{5}{4} = \frac{-10}{-8}$ and $5 > 4$, but -10 is *not* greater than -8. Therefore, here $c < d$.

Chapter 5

Geometric Gems

The origin of geometry lies, on the one hand, in problems of field measurement, and, on the other hand, in astronomy. As the field of mathematics evolved, one of the early formalities was Euclid's *Elements* (300 BCE), where we find the first attempt to introduce a system of axioms and postulates to establishing theorems. In a sense, geometry was the first exact and accurate science in which logic (formal) reasoning was required! Furthermore, geometry is a very concrete and demonstrative discipline; one can see and examine figures, shapes, solids, and patterns. Many students learn to really appreciate geometry — especially when it is well taught — and end up preferring it over arithmetic and algebra, largely because of its visual aspects.

Some Important Aspects of Geometry

Forms, figures, shapes, and patterns

The very graphic, concrete, clear, and vivid aspect of geometry presented in school offers many people something visually fascinating. Later in life, geometry helps define patterns and symmetry.

Geometry serves as a model for perceiving and experiencing space and objects around us

This includes everyday objects, such as mirrors, balls, playing fields, shapes of houses, crystals, and honeycombs.

Building a logic system using rigor with axioms, postulates, and proved theorems

Our intuition should be controlled by more exact considerations because otherwise we are tempted to draw wrong conclusions, such as that for a rectangle, where a false conclusion such as that the length of the diagonal equals the sum of the rectangle's length and width. This faulty statement evolves from the following "reasoning." In Figure 5.1, the sum of the steps along diagonal d is equal to $a + b$, since the sum of the steps is actually the sum of the horizontal parts and vertical parts. The same is true for Figure 5.2. On the one hand, the sum of the lengths of all stairs from A to C stays constant $a + b$ (even in case of arbitrarily small stairs). On the other hand, the whole stairway comes arbitrarily near to the diagonal when the number of stairs increases and the stairs become arbitrarily small, and we are led to the result that $d = a + b$, which we know is incorrect as it is contrary to the famous Pythagorean theorem which states that $a^2 + b^2 = d^2$. We cannot clarify here the deeper details where infinity plays a curious role.

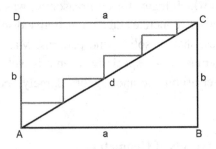

Figure 5.1

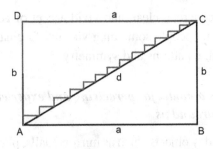

Figure 5.2

To truly appreciate the field of geometry, we now offer a collection of geometry problems that show the power and beauty of geometry with its unusual aspects.

Intersection Angle on a Clock

Problem: On the circular dial of a clock, the points at 2 and 8 are joined by a straight line, and the points at 6 and 11 are also joined by a straight line. These 2 lines form the angle φ. The challenge here is to determine the size of this angle φ (see Figure 5.3).

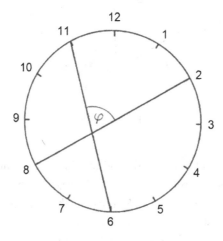

Figure 5.3

Solution: There are several ways to approach this challenge. We will consider three ways here.

(1) In Figure 5.4, we draw a parallel line to the line segment joining 6 and 11 and containing point M, which is the center of the circular clock. With points P and Q as the midpoints between with 5–6 and 11–12, respectively, we construct PQ containing the center of the clock at point M and it is parallel to line 11S6 as shown in Figure 5.4. Since the angle measured between numbers on the face of the clock is 30°, we then have the central angle $\angle 8MP = 75°$. However, since we have alternate

interior angles of parallel lines, we can conclude that $\varphi = \angle 8MP = 75°$.

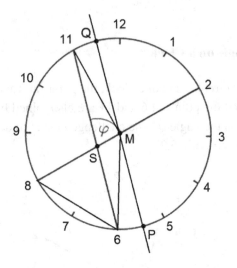

Figure 5.4

(2) We know that $\angle 11M\,8 = 90°$ and arc $118 = 90°$ and, therefore, the inscribed angle $\angle 1168 = 45° = \angle S\,68$; since $\Delta 86\,M$ is equilateral we know $\angle 68\,M = 60° = \angle 68\,S$; thus, in the triangle $\Delta S\,86$ we know $\varphi = \angle 8S\,6 = 180° - (60° + 45°) = 75°$.

(3) Another method begins with $\angle 6M\,11 = 90° + 60° = 150°$, that means $\angle M\,11\,S = 15°$ (note that the triangle $\Delta M\,116$ is isosceles) and, therefore, in $\Delta S\,M\,11$ we have $\varphi = \angle M\,S\,11 = 180° - (90° + 15°) = 75°$.

Angle Sum in Star Polygons

Problem: The challenge is to determine if there is a constant sum of the angle measures at the vertices of a pentagram (a star with five vertices) or of a heptagram (a star with seven vertices), which are shown in Figure 5.5.

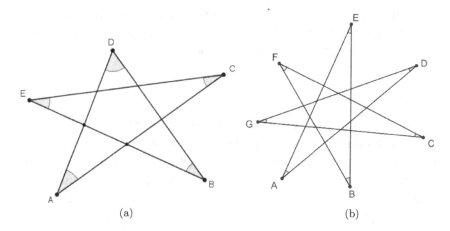

Figure 5.5

Is the angle sum the same for all such figures? If yes, what is the respective angle sum? If not, give examples with different angle sums.

Pentagram solution

As shown in Figure 5.6, we let α, β, and γ be the interior angles of the $\triangle AGF$, whereby $\alpha + \beta + \gamma = 180°$. For the corresponding exterior angles, we have $\angle CGE = 180° - \alpha$ and $\angle BFD = 180° - \beta$.

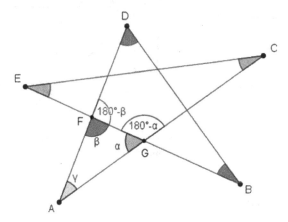

Figure 5.6

In the triangle $\triangle GCE$, we have $\angle GEC + \angle GCE = \alpha$. Analogously, in $\triangle FBD$, we have $\angle FDB + \angle FBD = \beta$. Thus, the angle sum of the pentagram is $\alpha + \beta + \gamma = 180°$, as we have noted above.

Heptagram solution

Again, we have an angle sum of $180°$. The reasoning is essentially the same as above in the case of a pentagram. One can see that in Figure 5.7 the exterior angle of triangle QDG is $\angle DQC = \theta + \delta$ and for triangle JEA the exterior $\angle DJE = \alpha + \varepsilon$. Then for triangle PHJ, the exterior angle $\angle QPC = \alpha + \varepsilon + \varphi + \beta$, and for triangle PQC, the sum of the angles $180° = \angle PCQ + \angle DQC + \angle QPC = \gamma + (\theta + \delta) + (\alpha + \varepsilon + \varphi + \beta)$.

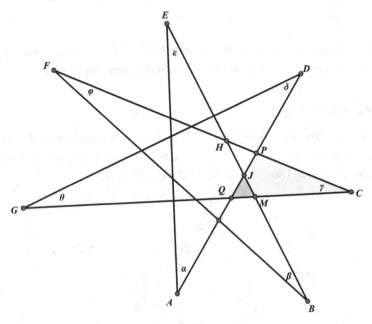

Figure 5.7

What is the Constant Angle?

Problem: In Figure 5.8, we show two congruent equilateral triangles that share a common vertex at point C. Their other vertices are joined by two lines forming angle *BPE*. Our challenge here is to find the measure of angle *BPE*.

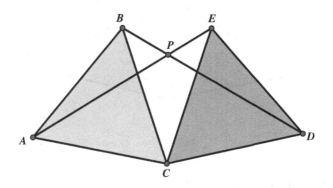

Figure 5.8

Solution: In Figure 5.9, we show two isosceles triangles, namely, $\triangle ACE$ and $\triangle DCB$, which generate equal base angles, $\angle EAC = \angle AEC$ and $\angle BDC = \angle DBC$. However, triangle *BCE* is also isosceles. It is not obvious that angle *BPE* does not depend on how far apart the two "upper" vertices *B* and *E* are. We have two congruent isosceles triangles $\triangle ACE$ and $\triangle DCB$, as their legs are sides of the congruent equilateral triangles, and the angles between them are equal: $\angle AEC = \angle DBC$. The angle whose measure we seek is $\angle BPE$, which is an exterior angle of $\triangle APB$ and, therefore, is the sum of the not adjacent interior angles:

$$\angle BPE = \angle BAP + \angle ABP = \angle BAP + (60° + \angle DBC)$$
$$= \angle BAP + (60° + \angle CAE) = 60° + \angle BAP + \angle CAE$$
$$= 60° + 60° = 120°$$

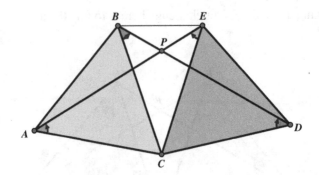

Figure 5.9

Strange Similar Triangles

Problem: In Figure 5.10, the point D on side BC of $\triangle ABC$ divides BC in the ratio $BD : DC = 1{:}2$. Furthermore, we are given the measures of two angles: $\angle CBA = 45°$ and $\angle CDA = 60°$. Our challenge is to prove that the triangles $\triangle ABC$ and $\triangle ADC$ are similar.

Hint: Draw the perpendicular from C to AD, which will facilitate a proof.

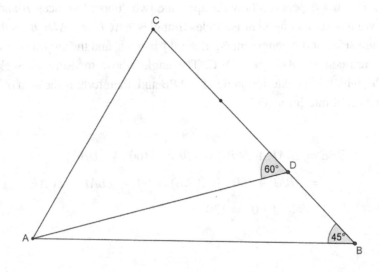

Figure 5.10

Solution: As suggested, we will draw the perpendicular from C to AD with the intersection point E, as shown in Figure 5.11. Then $\triangle EDC$ is a $30°-60°-90°$ triangle, and, therefore, we have $CD = 2ED$ so that, consequently, we have $ED = BD$. From this it follows that $\triangle EBD$ is isosceles with base angles of $30°$. Hence, also the triangle $\triangle EBC$ is isosceles with base angles of $30°$, and therefore, triangle $\triangle ABE$ is isosceles since it has equal base angles of $15°$. Thus, we can conclude that $AE = EC$. The isosceles right triangle $\triangle AEC$ has base angles of $45°$. Thus, we have two similar triangles $\triangle ABC$ and $\triangle ADC$ as they both have angles of $45°$, $60°$, and $75°$.

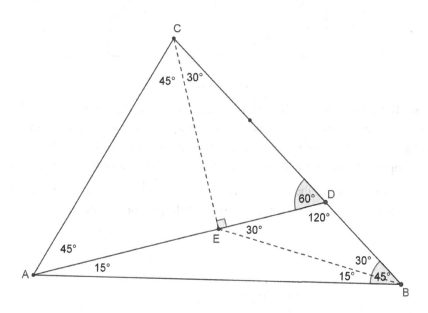

Figure 5.11

The Semicircle in a Quarter Circle

Problem: A semicircle is inscribed in a quarter circle as shown in Figure 5.12. We need to determine what portion of the quarter circle is the area of the semicircle.

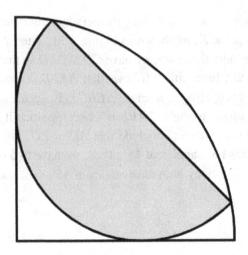

Figure 5.12

Solution: Let r denote the radius of the semicircle shown in Figure 5.13. Then we can draw a square with side r and diagonal $r\sqrt{2}$. The right triangle AFG with legs r and $r\sqrt{2}$ has hypotenuse $AG = R$, which by the Pythagorean theorem is $R = r\sqrt{3}$. Therefore, the area of the semicircle is $\pi \frac{r^2}{2}$, and the area of the quarter circle is $\pi \frac{R^2}{4} = \pi \frac{(r\sqrt{3})^2}{4} = \pi \frac{3r^2}{4}$. Thus, we have the fraction representing the semicircle part of the quarter circle as

$$\frac{\pi \frac{r^2}{2}}{\pi \frac{3r^2}{4}} = \frac{\frac{1}{2}}{\frac{3}{4}} = \frac{2}{3}.$$

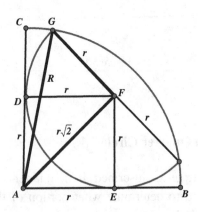

Figure 5.13

Determine an Unusual Angle Relationship

Here we are faced with a curious conundrum. In Figure 5.14, a semicircle with center M and diameter AB has a chord CD drawn parallel to AB. When we know that $\angle BAC = \alpha = \angle CMD$, what is the measure of α?

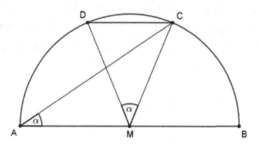

Figure 5.14

Solution: Since angle CAB is inscribed in arc CB, we have arc $BC = 2\alpha$, and because $CD \| AB$, we also know that arc $AD = 2\alpha$. Therefore, $CD = 180° - 4\alpha$. Thus, for the central $\angle DMC$ we know $\alpha = 180° - 4\alpha$ and $\alpha = \frac{180°}{5} = 36°$.

Discovering the Missing Area

A rectangle $ABCD$ is partitioned into four triangles by an interior point P. Three areas are given as shown in Figure 5.15. Our challenge here is to find the area of the fourth $\triangle APD$.

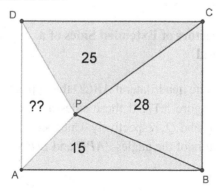

Figure 5.15

Solution: One method for determining the area of triangle *APD* is to draw the parallel line *EF* to *AB* through *P*, as shown in Figure 5.16. Since triangle *APB* and rectangle *APFE* have the same base and altitude, we know that the area of rectangle *ABFE* is twice the area of triangle *APB* or 30. Analogously, the area of rectangle *CDEF* is twice the area of triangle *CDP* or 50. Thus, the area of the rectangle *ABCD* is $50 + 30 = 80$, and by subtracting the sum of the three given triangle areas, $25 + 28 + 15 = 68$ from the area of the rectangle *ABCD* we find that the missing area, that of triangle *APD*, is 12.

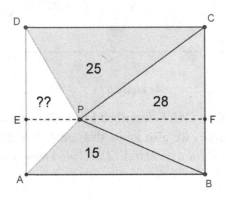

Figure 5.16

It should be noted that the above problem could also have been presented as a parallelogram *ABCD* rather than a rectangle *ABCD*, and the method of solution would have been analogous to the one presented here.

Perpendicular Bisectors of Extended Sides of a Cyclic Quadrilateral

Assume that in a cyclic quadrilateral *ABCD* the opposite sides are not parallel, as shown in Figure 5.17. If these sides are extended, one gets the intersection points *P* and *Q*, respectively. Unexpectedly, we need to justify that the angle bisectors of the angles $\angle APD$ and $\angle DQC$ are perpendicular.

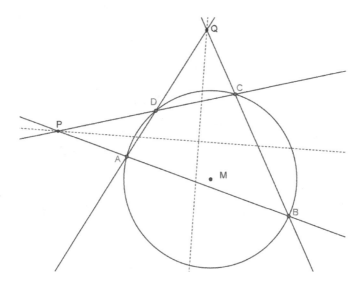

Figure 5.17

Solution: As we can see in Figure 5.18, the interior angles $\angle DAB$ and $\angle ABC$ of the cyclic quadrilateral are denoted as α and β, respectively. Since the opposite angles of a cyclic quadrilateral are supplementary, we get $\angle ADC = 180° - \beta$ and $\angle CDQ = \beta = \angle ADP$. From that we can conclude that in triangle ADP, we have $\angle APD = 180° - (180° - \alpha + \beta) = \alpha - \beta$, whereupon, because of the angle bisector, we have $\angle EPD = \frac{\alpha - \beta}{2}$. We also see in $\triangle ABQ$ we have $\angle AQE = \frac{180° - (\alpha + \beta)}{2} = 90° - \frac{\alpha + \beta}{2}$. And because the angle sum of a quadrilateral (also in concave ones such as $PEQD$) is $360°$, we can determine that $\angle PEQ = 360° - (180° + \beta) - \frac{\alpha - \beta}{2} - \left(90° - \frac{\alpha + \beta}{2}\right) = 90°$, which is what we had to show.

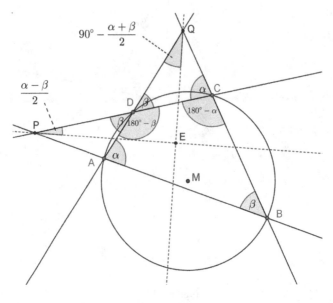

Figure 5.18

A Curious Square Consisting of Eight Rectangles with Equal Areas

Square *ABCD* consists of eight rectangles of equal area, as is shown in Figure 5.19. If the width of the shaded rectangle is 35, we need to find the area of the square *ABCD*.

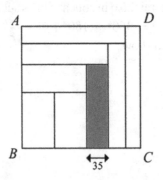

Figure 5.19

Solution: Here algebra is a helpful tool to find the sought-after area. As shown in Figure 5.20, we need to get the values of a, f, and x, which will then provide the length of a side of the square and subsequently the area of the square.

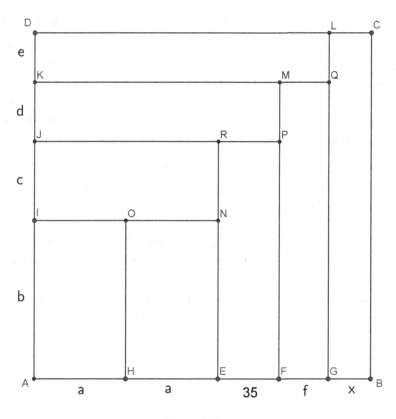

Figure 5.20

In Figure 5.20, the various lengths are labeled a, b, c, d, e, f, x. Now, we will generate the steps to determine the values of the crucial values a, f, and x, which will determine the length of the side of the square. We begin by recognizing that the area $AHOI$ = area $INRJ$, which gives us $ab = 2ac$, and, thereby, $b = 2c$. Then since area $EFPR$ = area $INRJ$, we can conclude that $35 \times \underbrace{(b+c)}_{3c} = 2ac$, which then yields $2a = 105$.

Now consider that the area $INRJ$ = area $JPMK$, which yields $105c = 140d$ and $d = \frac{3}{4}c$. Next, we will consider that area $FGQM$ = area $JPMK$ and then get $f\underbrace{\left(3 + \frac{3}{4}\right)c}_{\frac{15}{4}} = 140(\underbrace{\frac{3}{4}c}_{d})$, which gives us $f = 28$. We then

use area $KQLD$ = area $FGQM$ to get $168e = \left(\frac{15}{4}c\right)28$, which results in $e = \frac{5}{8}c$. Finally, we use area $GBCL$ = area $KQLD$ and conclude $x\left(\frac{15}{4} + \frac{5}{8}\right)c = 168\left(\frac{5}{8}c\right)$, thus, $x = 24$. We then have the bottom side length of the square as $a + a + 35 + f + x = 105 + 35 + 28 + 24 = 192$. Therefore, the area of the square is $192^2 = 36,864$. We could have found the area of the square by obtaining the values b, c, d, e, however, this would have been more difficult since these would not have been integer values.

Comparison of Circular Areas

In Figure 5.21, we show a circle and semicircle with the same center point at O. The point B of the semicircle is the intersection point of the tangents to the circle at A and C, respectively. Our challenge here is to determine how the area of the circle compares to the area of the (bold) semicircle.

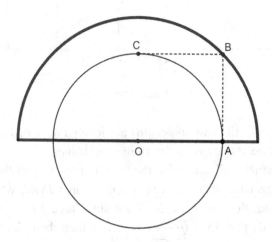

Figure 5.21

Solution: Let us begin by assuming that the radius of the circle is 1. With the perpendiculars at points A, O, and C, we have quadrilateral $ABCO$ as a rectangle. However, since tangents AB and CB are equal, we know that $ABCO$ is a square. Then in Figure 5.22, by the Pythagorean theorem, since $AO = OC = AB = 1$, we get $OB = \sqrt{2}$, which is the radius of the bold semicircle. Thus, the area of the circle is $\pi (AO)^2 = \pi$, and the area of the (bold) semicircle is $\frac{\pi (OB)^2}{2} = \frac{\pi \left(\sqrt{2}\right)^2}{2} = \pi$. Quite unexpectedly, we find that the two areas are equal.

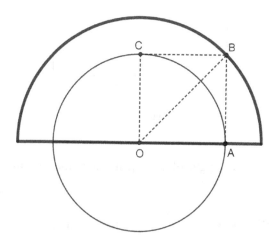

Figure 5.22

Comparison of a Circular Area with a Square

In Figure 5.23, the square and the circle share a common center point, P. The line $PQSR = PA$ and is parallel to the side of the square. Also point Q is on a side of the square, while point S is on the circle and $SR = 2SQ$. We now need to find how the area of the circle compares to the area of the square.

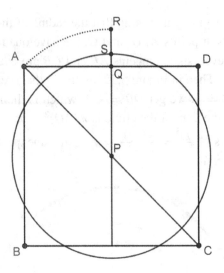

Figure 5.23

Solution: If the side length of the square is 1, then by the Pythagorean theorem

$$AP = \frac{\sqrt{2}}{2} = PR, \text{ and } RQ = \frac{\sqrt{2}}{2} - \frac{1}{2}, \text{ whereupon, } SQ = \frac{1}{3}\left(\frac{\sqrt{2}}{2} - \frac{1}{2}\right).$$

Then

$$PS = PQ + PS = \frac{1}{2} + \frac{1}{3}\left(\frac{\sqrt{2}}{2} - \frac{1}{2}\right) = \frac{2 + \sqrt{2}}{6}.$$

Thus, the area of the circle is $\pi \cdot \frac{(2+\sqrt{2})^2}{36} \approx 1.017$. The area of the square is simply 1, therefore, the circle is about 1.7% larger than the square.

What Is the Area of the Circle?

In Figure 5.24, we have two squares $ABCD$ and $DEFG$ with areas 18 and 4, respectively. The two squares have a common vertex at point D, and between them $\angle CDG = 45°$. We need to find the area of the circumscribed circle.

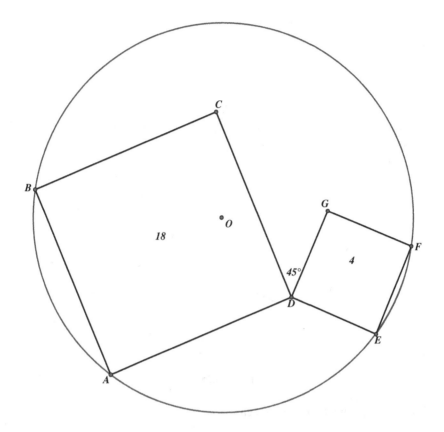

Figure 5.24

Solution: In Figure 5.25, we find that $\angle FDG = 45° = \angle CDG$, which means that together they form a right-angle FDC. This allows us to conclude that ADF is a straight line and, therefore, we find that $\triangle BAF$ is a right triangle. Thus, BOF is the diameter of the circle. Now, since the area of square $DEFG$ is 4, we have $GF = 2$ and applying the Pythagorean theorem to triangle DEF, we find that $DF = 2\sqrt{2}$. Since AB and AD are sides of square $ABCD$ whose area is 18, we get $AB = AD = \sqrt{18} = 3\sqrt{2}$. Applying the Pytharorean theorem to triangle ABF, we find that the diameter $BF = \sqrt{(3\sqrt{2})^2 + (3\sqrt{2} + 2\sqrt{2})^2} = \sqrt{18 + 50} = \sqrt{68} = 2\sqrt{17}$. Thus, the radius of the circle is $\sqrt{17}$, and the area of the circumscribed circle is $\pi(\sqrt{17})^2 = 17\pi \approx 53.4$.

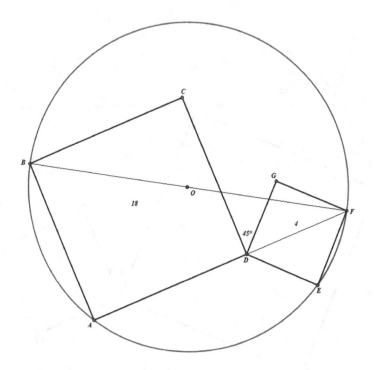

Figure 5.25

Comparing Related Circle Areas

An equilateral triangle and a regular hexagon have the same perimeter, as shown in Figure 5.26. We need to determine the ratio of the areas of their inscribed circles.

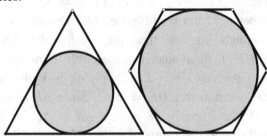

Figure 5.26

Solution: One can answer this question rather quickly without any significant calculations. We must realize that the two polygons, the triangle and the hexagon, have the same height as shown in Figure 5.26. Therefore, half the height, h, would be the radius of the circle inscribed in the equilateral hexagon, yielding an area of $\pi \left(\frac{h}{2}\right)^2 = \frac{\pi h^2}{4}$. The center of the inscribed circle in the equilateral triangle is at the point of intersection of the medians, which is a trisection point for each of the medians, as in this case the medians are also the angle bisectors, which determine the center of the inscribed circle. This then implies that the radius of the circle inscribed in the equilateral triangle is $\frac{h}{3}$. Thus, the area of the triangle's inscribed circle is equal to $\pi \left(\frac{h}{3}\right)^2 = \frac{\pi h^2}{9}$. Thus, the ratio of the corresponding areas is 4:9. We could have come to this response more quickly by considering that since the ratio of the radii of the two circles is 2:3, then the ratio of the areas is 4:9.

Another Comparison of Areas

In Figure 5.27, a semicircle is placed inside a square so that the diameter of the semicircle forms 45° angles with two sides of the square and the other two sides of the square are tangent to the semicircle. The challenge here is to determine how the area of the semicircle (the shaded region) compares to the non-shaded regions in the square.

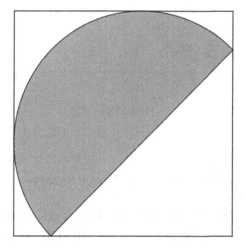

Figure 5.27

Solution: Let us denote the radius of the semicircle by r, so its area is $\frac{\pi}{2}r^2 \approx$ $1.57r^2$. The side length of a square with diagonal r is $\frac{r}{\sqrt{2}}$. Therefore, the area of the unshaded isosceles right triangle EBH, as shown in Figure 5.28, is $2 \cdot \left(\frac{r}{\sqrt{2}}\right)^2 = r^2$. The upper left unshaded "corner" (KDJ) of the square can be found by taking one-quarter of the circle from which the semicircle came and subtracting it from the square $KMJD$. That gives us $r^2 - \frac{\pi}{4}r^2$. Each of the two other unshaded regions near the corners A and C, which are congruent, has an area of $r\frac{r}{\sqrt{2}} - \frac{r^2}{4} - \frac{\pi}{8}r^2 = \frac{4\sqrt{2}-2-\pi}{8}r^2$, which is obtained by $MG = \frac{r}{\sqrt{2}} = MF$. The sum of the non-shaded regions, therefore, is $\left(2 - \frac{\pi}{4} + 2\frac{4\sqrt{2}-2-\pi}{8}\right)r^2 = \frac{3+2\sqrt{2}-\pi}{2}r^2 \approx 1.34r^2$. Thus, the ratio of the areas of the shaded to the non-shaded region is approximately $1.57:1.34 \approx 1.17:1$.

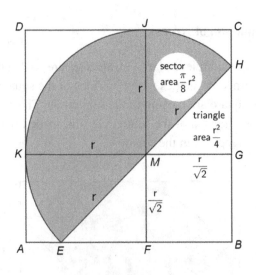

Figure 5.28

Finding the Area of an Awkwardly Placed Square

In Figure 5.29, the square $ABCD$ is placed on a circle so that vertices C and D lie on the circle which is tangent to side AB of the square. Line segment EF, whose length is 1, is the perpendicular bisector of DC at point E and intersects the circle at point F. The challenge here is to find the area of square $ABCD$.

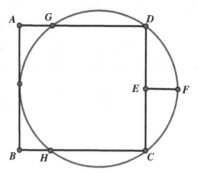

Figure 5.29

Solution: Let a represent the side length of the square and let r represent the radius of the circle. If we imagine that the square had slid a distance of EF away from the right side of the circle (where, before moving, due to symmetry reasons it had been tangent to the circle at F and the left vertices had been at G and H) so that it is tangent to the left side of the circle, we can conclude that BH must indicate the distance moved, so that $BH = EF = 1$. We then have $GD = AD - AG = a - 1$. In Figure 5.30, we see that angle GDC is a right angle, therefore, GC must be the diameter of the circle. Then, by the Pythagorean theorem applied to triangle GDC, we have $(2r)^2 = (a - 1)^2 + a^2$; if we insert $2r = a + 1$, we get $a = 4$, which yields the area of the square $a^2 = 16$.

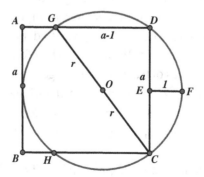

Figure 5.30

Comparison of Perimeters

In Figure 5.31, we show a square with two of its vertices on the circle, and the opposite side tangent to the circle. Our task here is to determine how the two perimeters compare. Furthermore, we also would like to know the fraction of the vertical edges of the square which lie outside the circle.

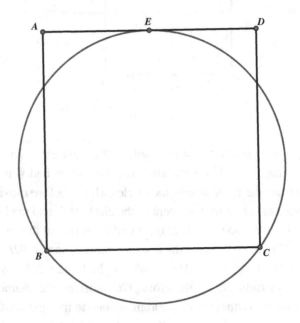

Figure 5.31

Solution: We denote half of the side length of the square with a, and the circle's radius with r, as we are shown in Figure 5.32. Applying the Pythagorean theorem to triangle BOF, we find that $OF = \sqrt{r^2 - a^2}$. However, we can also represent OF as $2a - r$. Therefore, $\sqrt{r^2 - a^2} = 2a - r$, and by squaring both sides of the equation, we get $r^2 - a^2 = 4a^2 - 4ar + r^2$, which then gives us $r = \frac{5}{4}a$. Thus, the perimeter of the square is $8a$, and the circumference of the circle is $2\pi r = 2\pi \frac{5}{4}a = \frac{5\pi}{2}a \approx 7.854a$. So, we can see that the perimeter of the square is slightly longer than the circumference of the circle.

    The vertical overlap is $2a - 2\sqrt{r^2 - a^2} = 2a - 2\sqrt{\frac{25}{16}a^2 - a^2} = \frac{a}{2}$, thus, the overlap is exactly one-fourth of the square's side length.

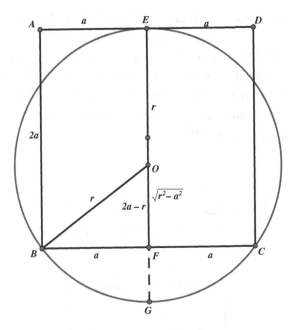

Figure 5.32

The Unexpected Region Evolving from Two Overlapping Quarter Circles

In Figure 5.33, there are two-quarter circles in a square. The challenge here is to determine the fraction of the square that is indicated by the shaded region determined by points *ADF*.

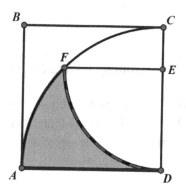

Figure 5.33

Solution: Let the side length of the square be 1. In Figure 5.34, we see that triangle *FED* is an isosceles right triangle and $\angle FDE = 45°$. Therefore, *DF* extended to point *B* becomes the diagonal of square *ABCD*. Also, $AD = FD = 1$, which is the hypotenuse of triangle *FED*. Therefore, applying the Pythagorean theorem to isosceles right triangle *DEF*, the small quarter circle has radius $FE = \frac{1}{\sqrt{2}}$. The area by the shaded region bounded by *ADF* can be determined by

$$\underbrace{\frac{\pi}{8}}_{\substack{\text{half of the big} \\ \text{quarter circle}}} - \left(\underbrace{\frac{\frac{1}{2} \cdot \pi}{4}}_{\substack{\text{small} \\ \text{quartercircle}}} - \underbrace{\frac{\frac{1}{\sqrt{2}} \cdot \frac{1}{\sqrt{2}}}{2}}_{\substack{\text{isosceles} \\ \text{right triangle } FED}} \right) = \frac{\pi}{8} - \left(\frac{\pi}{8} - \frac{1}{4} \right) = \frac{1}{4}.$$

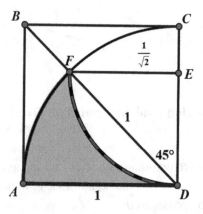

Figure 5.34

Comparing the Area of Seemingly Unrelated Triangles

The semicircle on *AB* is intersected by a circle at points *P* and *R*, which is shown in Figure 5.35. Line *AP* extended intersects the circle at point *D*, and the line *BR* extended intersects the circle at point *C*. Finally, *AC* and

BD intersect at *E*. The challenge here is to find the relationship between the areas of triangles *ADE* and *BCE*.

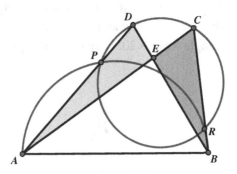

Figure 5.35

Solution: In Figure 5.36, when we draw *PR* and *CD*, we note that there are two cyclic quadrilaterals: *ABRP* and *PRCD*. This allows us to generate the following: $180° - \angle BAP \overset{ABRP}{\underset{\text{is cyclic}}{=}} \angle PRB \overset{PRCD}{\underset{\text{is cyclic}}{=}} \angle PDC = \angle ADC$ and therefore, supplementary angles *BAD* and *ADC* enable us to conclude that *AB*||*CD*. From this, it follows that the triangles $\triangle ACD$ and $\triangle BCD$ have equal areas since they share the same base *CD* and have equal altitudes. If we now subtract the common area of $\triangle ECD$ from these two equal-area triangles, we are left with the area of triangle *ADE* equal to the area of triangle *BCE*.

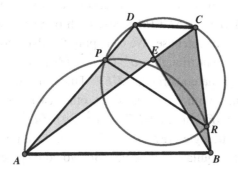

Figure 5.36

Four Equilateral Triangles and Two Given Areas

In Figure 5.37, we have four equilateral triangles: $\triangle ABC$, $\triangle CDE$, $\triangle HEF$, and $\triangle AFG$. Triangle *HEF* is placed to that its sides are perpendicular to the sides of triangle ABC. We know that know that the area of triangle *CDE* is 5, and the area of triangle *AFG* is 20. Here the challenge is to find the area of triangle *HEF*.

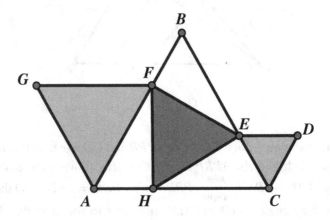

Figure 5.37

Solution: Since the area of triangle *AFG* is four times the area of triangle *CDE*, the sides are in a ratio of 1:2 so that we can let $EC = a$ and $AF = 2a$, and their areas are in a ratio of $a^2 : 4a^2$. We can also show that $\triangle AHF \cong \triangle BFE \cong \triangle CEH$ because each one has a 60° angle, a 90° angle and an equal hypotenuse. We can apply the Pythagorean theorem to one of these congruent triangles: $a^2 + x^2 = (2a)^2$ and $x = a\sqrt{3}$, which is a side of the equilateral triangle *HEF*. Therefore, the area of triangle *HEF* is in the ratio with the areas of the other two triangles as follows: area$\triangle CDE$: area$\triangle AFG$: area$\triangle HEF = a^2 : (2a)^2 : (a\sqrt{3})^2 = a^2 : 4a^2 : 3a^2$. Thus, the area of triangle *HEF* is $3 \times 5 = 15$. As a bonus, we can find that the area of the equilateral triangle *ABC* would then be $9 \times 5 = 45$.

Five Congruent Triangles in a Rectangle

In Figure 5.38, we show five congruent right triangles fitted into a rectangle. The challenge here is to find what fraction of the rectangle's area is covered by the shaded triangles.

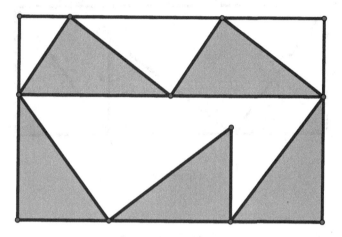

Figure 5.38

Solution: As we show in Figure 5.39, we can represent the sides of one of the congruent right triangles as $a \leq b \leq c$, so that by the Pythagorean theorem we get $a^2 + b^2 = c^2$. Furthermore, we see that the length of the rectangle can be represented in two ways, which gives us the following equation: $2c = 2a + b$ then $b = 2(c - a)$. If we insert $b = 2(c - a)$ into the Pythagoras equation and solve for c ($c \geq a$), we get $c = \frac{5}{3}a$, and if we then substitute that into the value of b, we get $b = 2\left(\frac{5}{3}a - a\right) = \frac{4}{3}a$. This indicates the ratio $\frac{b}{a} = \frac{4}{3}$, which leads us to the notion that we have a 3-4-5 right triangle. The area of one of these congruent right triangles is $\frac{1}{2}ab = (\frac{1}{2})(3)(4) = 6$, and with h as the altitude to the hypotenuse of one of the five congruent triangles, this also can be expressed by $\frac{1}{2}ch = \frac{1}{2}(5h) = 6$ so that the altitude $h = 2.4$. Thus, the ratio of the sum of the areas of the five right triangles to the area of the rectangle is $\frac{5 \times \frac{3 \times 4}{2}}{10 \times 6.4} = \frac{30}{64} = \frac{15}{32}$.

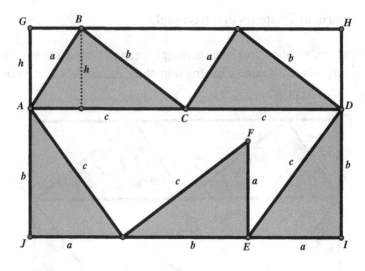

Figure 5.39

Two Overlapping Congruent Right Triangles

In Figure 5.40, we show two congruent overlapping right triangles. Our challenge is to find the fraction of the entire shaded figure that is covered by the overlapping triangles.

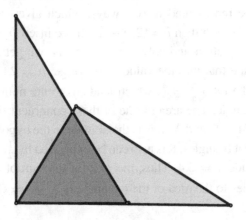

Figure 5.40

Solution: In Figure 5.41, triangles ABC and DEB are congruent with an area of X. Since $\angle ACB = \angle DBE$ we know that the darker-shaded triangle BFC is isosceles so that $FC = FB$. Because $\angle FBA = 90° - \angle DBE$ and $\angle BAF = 90° - \angle ACB$, we have $\angle FBA = \angle BAF$, and, thus, $\triangle ABF$ is isosceles. It follows that $FA = FB$ and then $FC = FA$. Therefore, we have triangle AFB equal in area to triangle CFB, since they both have the same altitude from B to CFA and have equal bases. This then tells us that the area of the triangle BFC is $\frac{X}{2}$. The complete shaded area is $2X - \frac{X}{2} = \frac{3X}{2}$, and thus, the sought-after fraction is $\frac{\frac{X}{2}}{\frac{3X}{2}} = \frac{1}{3}$.

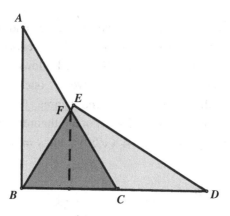

Figure 5.41

Discovering a Square's Area

Square $ABCD$ has its opposite vertices A and C connected with three lines with perpendicularity shown as in Figure 5.42 and lengths $CF = 3$, $EF = 1$, and $AE = 4$. Our challenge here is to find the area of square $ABCD$.

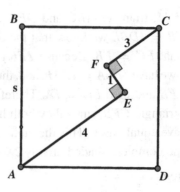

Figure 5.42

Solution: As we show in Figure 5.43, we begin by extending AE to intersect at point G the perpendicular from point C to AE extended. From rectangle $CGEF$, we find that $CG = 1$ and $AG = 7$. Following the Pythagorean theorem, we find that $AC = \sqrt{50}$. This gives us one aspect of the size of the square, which will then allow us to find its area. The side of the square $ABCD$ is then easily found by the Pythagorean theorem applied to triangle ABC, where $2s^2 = AC^2$, then $2s^2 = (\sqrt{50})^2$, and $s = 5$, and this yields an area of 25.

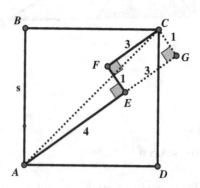

Figure 5.43

Squares on the Sides of a Parallelogram

On the sides of the parallelogram *ABCD*, squares are erected, as shown in Figure 5.44. The centers of the four squares are *E*, *F*, *G*, and *H*. Curiously, regardless of the shape of the parallelogram, these four points *E*, *F*, *G*, and *H* turn out to be the vertices of a square. We need to show how this can be proved.

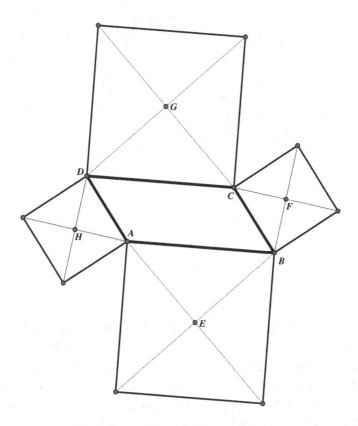

Figure 5.44

Solution: Consider the four triangles $\triangle HAE$, $\triangle HDG$, $\triangle EBF$, and $\triangle FCG$ shown in Figure 5.45. We can show that these four triangles are all congruent by simply choosing two of these triangles to prove their congruence and then realizing that the procedure can be followed for the other two as well.

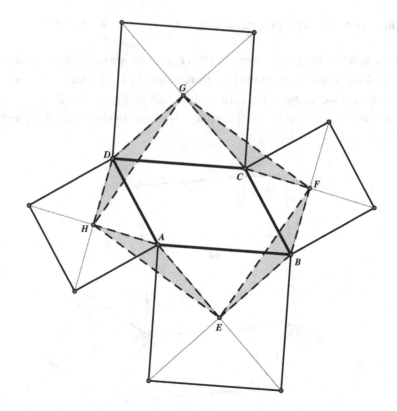

Figure 5.45

Consider the two triangles $\triangle FBE$ and $\triangle HDG$. We have many 45° angles in the diagram such as $\angle GDC = \angle HDA = \angle FBC = \angle EBC = 45°$, and we also know that $\angle ADC = \angle ABC$. Therefore, by addition, we find that $\angle GDH = \angle EBF$. Furthermore, $DH = BF$ and $GD = BE$. Thus, $\triangle FBE \cong \triangle HDG$ and $GH = FE$. Using this procedure, we can show that the four triangles are all congruent with one another. This makes quadrilateral $GHEF$ a rhombus. However, we can also show that each of its angles is a right angle, since $\angle DHG = \angle AHE$ and because $\angle DHG + AHG = 90°$, it follows that $\angle AHE + AHG = 90°$, which is one of the angles of the rhombus, thus, making it a square.

Unexpected Similar Triangles

On the sides of a triangle $\triangle XYZ$, similar triangles (with equal angles α, β, γ) are drawn outwardly so that each of the angles is at the outer vertex exactly once, as shown in Figure 5.46. The points P, Q, and R are the centers of the circumscribed circles of the external triangles. We need to show that the triangle $\triangle PQR$ is actually similar to the three similar triangles.

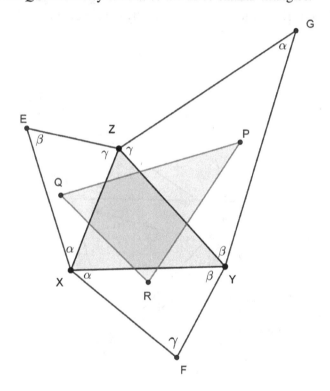

Figure 5.46

Solution: As shown in Figure 5.47, let A be the intersection point of the circumcircles p and q, then we have the supplementary opposite angles of cyclic quadrilaterals $ZAXE$ and $ZAYG$, $\angle ZAX = 180° - \beta$ and $\angle ZAY = 180° - \alpha$, respectively. Thus $\angle XAY = 360° - (\angle ZAX + \angle ZAY) = 360° - (180° - \beta + 180° - \alpha) = \alpha + \beta = 180° - \gamma$. And this, in turn, means that point A lies on the circumcircle r, and we have proved that the three circumcircles intersect each other at point A. Furthermore, the sides of triangle PQR, which are PQ, QR, and RP, are perpendicular to AZ, AX, and AY, respectively. As we had earlier, $\angle ZAY = 180° - \alpha$ and $\angle ZAY = \angle DAC$, thus, we have $\angle DAC = 180° - \alpha$. Therefore, since quadrilateral $DPAC$ is cyclic, because it has two opposite right angles, we then have $\angle P = \alpha$. This same procedure can be used for the other two angles of triangle PQR. Hence, for the interior angles of the triangle $\triangle PQR$, we get α, β, γ and thus, $\triangle PQR$ is similar to the three externally constructed triangles.

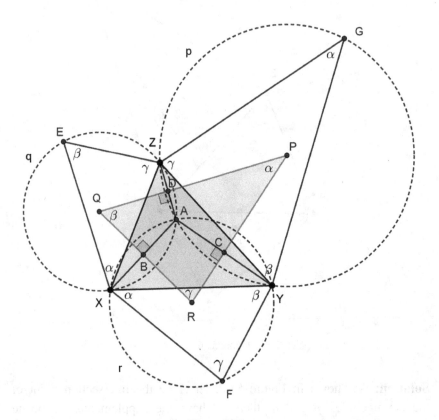

Figure 5.47

Tangential Quadrilateral

As we can see in Figure 5.48, the circle with center I is inscribed in quadrilateral *ABCD*. Perpendiculars *HE, EF, GF*, and *HG* are drawn to *AI, BI, CI*, and *DI*, respectively. These perpendiculars determine quadrilateral *EFGH*. Unexpectedly, we find that the diagonals *EG* and *FH* intersect at point *I*.

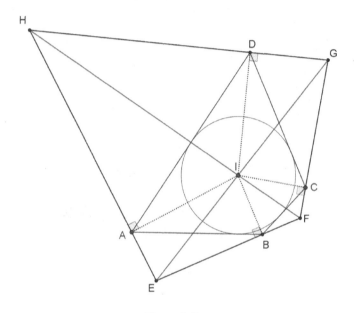

Figure 5.48

Solution: Since one pair of opposite angles of each of the quadrilaterals *AIBE, DICG, BICF*, and *HDIA* are right angles, each of these quadrilaterals is cyclic. Let the interior angles of the circumscribed quadrilateral *ABCD* be $2\alpha, 2\beta, 2\gamma, 2\delta$, as shown in Figure 5.49. Since inscribed angles measured by the same arc are equal, we can see how the angles at E, F, G, H are determined by $\alpha, \beta, \gamma, \delta$. As an example, consider the angles at point H: Since $\angle IAD = \alpha$ and in the circumcircle of the cyclic quadrilateral *AIDH*, the angle $\angle IHD$ is another inscribed angle measured by $\widehat{ID}$, we can conclude that $\angle IHD = \alpha$. And using the arc $\widehat{IA}$ in the same circle, we have $\angle ADI = \delta = \angle AHI$. Since the sum of the interior angles of *EFGH* is 360°, we know that $2\alpha + 2\beta + 2\gamma + 2\delta = 360°$, which implies that $\alpha + \beta + \gamma + \delta = 180°$, and this is the sum of the angles in triangle *HEF*. Hence, *EIG* is a straight line passing through point *I*, essentially indicating that the diagonal *EG* contains point *I*.

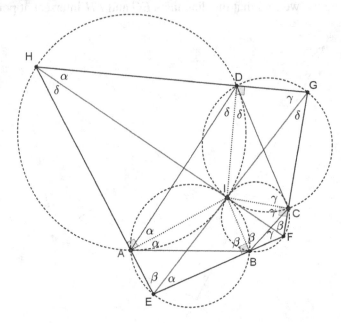

Figure 5.49

The Moving Semicircle

In Figure 5.50, a semicircle is placed with its diameter's endpoints on two perpendicular lines. A point *P* is placed somewhere on the arc of the semicircle. As the endpoints of the diameter move along the perpendicular axes, we need to determine what the path of the point *P* on the semicircle's arc is.

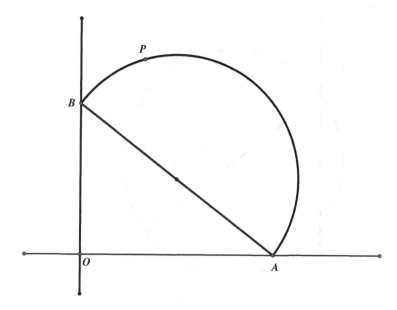

Figure 5.50

Solution: First, one must realize that when the semicircle is completed, it must pass through the perpendicular intersection, since the diameter of the circle generates a right angle, which happens to be point O. This can be seen in Figure 5.51. As the points B and A slide along the two axes, one constant will be $\angle POB = \angle PAB$ since they are both measured by arc PB. Therefore, the point P will always travel along the line PO.

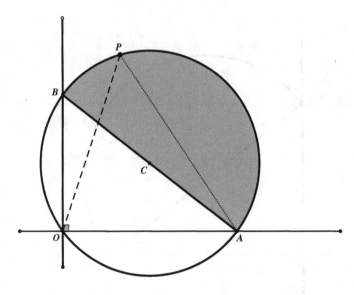

Figure 5.51

Shaded Area in a Rectangle

In Figure 5.52, we show a rectangle *ABCD*, with two congruent right trian-gles, *AFB* and *DEC*, whose right angles are at points *E* and *F*, which are on sides *AB* and *DC*, respectively. Our task at hand is to determine what fraction of the rectangle is shaded.

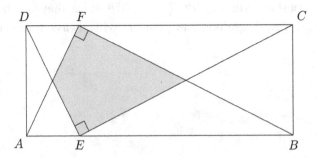

Figure 5.52

Solution: As we show in Figure 5.53, we draw the line segment *EF* and two rectangles *ADFE* and *BCFE* arise. We can easily see that one-fourth of each of the two rectangles' respective areas is shaded. This results in one-fourth of the total rectangle *ABCD* being shaded.

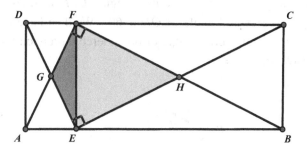

Figure 5.53

A Surprise Chord Length

On diameter *AB* of a semicircle with radius *r*, an arbitrary point *X* is chosen, and two lines are drawn from point *X* to the semicircle's arc, which makes three 60° angles with *AB*, as shown in Figure 5.54, and intersects the semicircle at points *C* and *D*. The challenge here is to determine the length *s* of chord *CD* in relation to the radius of the circle.

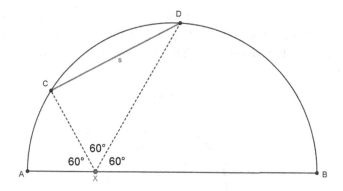

Figure 5.54

Solution: As we show in Figure 5.55, we draw a perpendicular to *AB* from point *D* which intersects the circle at point *F*. This produces two right triangles $\triangle XED$ and $\triangle XEF$ with angles of $60°$ and $30°$. Because of the three $60°$ angles at point X, one can see that F lies also on the extension of CX, and that yields the $30°$ inscribed angle $\angle CFD$ with arc CD. Thus, the corresponding central angle $\angle COD$ is $60°$. We then have an isosceles triangle CDO with the vertex angle of $60°$, thereby, producing equilateral triangle CDO, where $s = r$.

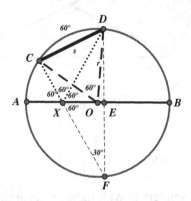

Figure 5.55

Area of a Curious Circle

Figure 5.56 shows four squares sharing a common vertex. The areas of three of these squares are shown to be 16, 25, and 100. A circle passes through one vertex of each of the squares. The challenge here is to find the area of the circle.

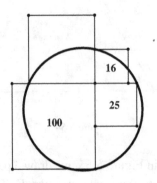

Figure 5.56

Solution: The side lengths of the three squares with given areas are 4, 5, and 10. By joining the points where the circle intersects the vertices of the squares, as we do in Figure 5.57, we find that, because the products of the segments of intersecting chords are equal, we have $a \times c = d \times b$ and $5a = 4 \times 10 = 40$ so that $a = 8$. Since DB and $A'C$ are perpendicular to AC, we have DB parallel to $A'C$. Therefore, the arc $A'B = $ arc CD. Thus, we get $A'B = DC = \sqrt{c^2 + d^2}$ from the Pythagorean theorem applied to triangle DEC. We then apply the Pythagorean theorem again, first to triangle AEB to get $AB = \sqrt{a^2 + b^2}$ and then to triangle $AA'B$ to find that $AA' = \sqrt{a^2 + b^2 + c^2 + d^2}$. Thus, the radius of the circle is $AM = \frac{\sqrt{a^2+b^2+c^2+d^2}}{2}$. The area of the circle is, therefore, $\pi \left(\frac{\sqrt{a^2+b^2+c^2+d^2}}{2} \right)^2 = \frac{\pi \left(a^2+b^2+c^2+d^2 \right)}{4}$, and for the given lengths, the area of the circle is $\frac{64+100+25+16}{4}\pi = \frac{205}{4}\pi \approx 161$.

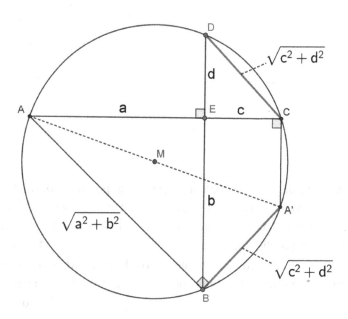

Figure 5.57

A Square on Two Parallel Lines

A square *ABCD* with side length *a* is placed onto two parallel lines whose distance apart is *a*, such that the square has one vertex on the bottom parallel, one vertex above the top parallel line, and two vertices between the parallel lines. We show two such cases in Figure 5.58. We need to show that in this situation the "overhanging" triangle $\triangle AEF$ has a constant perimeter regardless of the position of the square. Furthermore, we need to find the value of this perimeter.

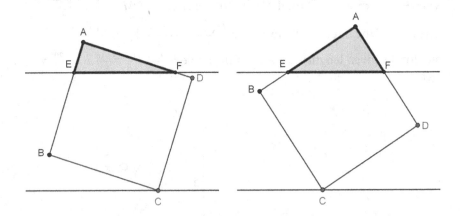

Figure 5.58

Solution: We can begin by conjecturing that the perimeter of triangle *AEF* is always 2*a*, which is twice the side length of the square. This value can be conjectured by considering an extreme position, where *A* gets arbitrarily close to the top parallel (or where we would find *A* = *E* or *A* = *F*). Then the triangle $\triangle AEF$ degenerates to the double line *EF*, where *EF* obviously moves to overlap the side length *a* of the square. But now let us prove this conjecture by showing that this holds for all positions of *A* above the top parallel.

Suppose the square *ABCD* is rotated around point *C*. We have to show that the perimeter of *AEF* does not depend on the position of point *A*, as shown in Figure 5.59. Draw a line perpendicular to the bottom parallel at point C and it intersects the top parallel line at point J.

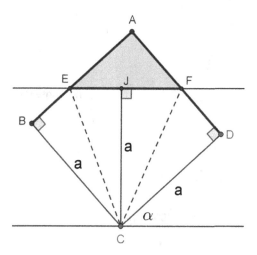

Figure 5.59

Now two quadrilaterals appear: $CDFJ$ and $CJEB$. Each has two congruent right triangles with equal legs and a common hypotenuse, so that triangles CBE and CJE are congruent, as are triangles CDF and CJF. Therefore, $EJ = EB$ and $JF = FD$, and one can see that the perimeter of triangle EFA is $AE + EJ + AF + FJ = AE + BE + AF + FJ = 2a$, which is twice the side length of the square.

Unexpected Equal Line Segments

Let $ABCDE$ be a pentagon with equal sides and two right angles at consecutive vertices C and D. Furthermore, we let P be the intersection point of the diagonals AC and BD. Our challenge here is to prove that $PA = PD$.

Solution: We intentionally did not provide a diagram following the statement of the problem, since making such a sketch is closely tied to the solution.

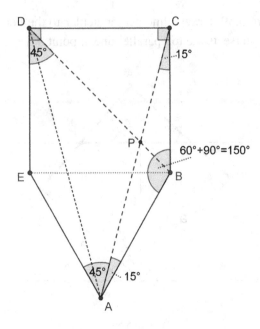

Figure 5.60

As shown in Figure 5.60, the pentagon has five equal sides and two right angles at consecutive vertices C and D; we can then conclude that $EBCD$ is a square and $\triangle ABE$ an equilateral triangle. Therefore, the angle at B, namely, $\angle ABC = 150°$, which establishes the base angles of isosceles triangle $\triangle ABC$ to be 15°. Therefore, $\angle EAC = 60° - 15° = 45°$. As the diagonal of the square bisects the angles, we have $\angle EDB = 45°$. Moreover, $\triangle AED$ is isosceles and has equal base angles, namely, $\angle EDA = \angle EAD$. Thus, by subtraction, $\angle PDA = \angle PAD$. Hence, we have isosceles $\triangle APD$ with $PA = PD$.

A Surprising Angle Sum

In Figure 5.61, we show two equal-area isosceles triangles $\triangle ABC$ and $\triangle ABD$, where $AB = CB$ and $AB = BD$. Our challenge here is to find the sum of $\angle ACB + \angle ADB$.

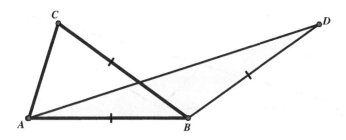

Figure 5.61

Solution: Since the area of triangle ABD is equal to the area of triangle ABC, and they share the same base, therefore, their altitudes must be equal, thus, making CD parallel to AB. Let $\alpha = \angle ACB$ and $\beta = \angle ADB$. Because of $\triangle ABC$ is isosceles it has equal base angles α, and then $\angle ABC = 180° - 2\alpha$. As we can see in Figure 5.62, the alternate-interior angles of the parallel lines, $\angle ADC = \angle BAD = \beta$. Therefore, the base angles of the isosceles triangle CBD are 2β. Once again, the alternate-interior angles of the parallel lines gives us $\angle ABC = \angle BCD$, which is that $180° - 2\alpha = 2\beta$, or $2\alpha + 2\beta = 180°$, which then provides us with our sought-after conclusion $\alpha + \beta = 90°$.

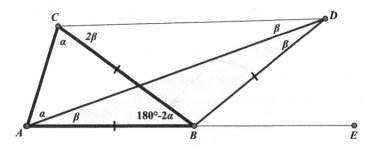

Figure 5.62

Determine the Angle

In Figure 5.63, we show two isosceles triangles, ABC and ABD, that have equal areas. We also show that $AB = BD$, $AC = BC$, and $\angle ACB = 90°$. Our challenge is to find the measure of angle ADB.

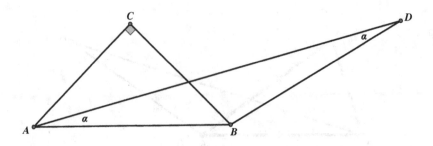

Figure 5.63

Solution 1: First, we know that the altitudes of the two isosceles triangles are equal since they share the same base, AB, and have equal areas. This implies that in Figure 5.64 $CD||AB$, and therefore, $\angle BAD = \angle ADC$, as they are alternate-interior angles of the parallel lines, thus $\angle ADC = \alpha$. If x denotes the length of each of the two legs in the right triangle $\triangle ABC$, we have $BD = AB = x\sqrt{2}$. With the help of the law of sines[2] applied to $\triangle BCD$, one gets $\frac{x\sqrt{2}}{\sin(45°)} = \frac{x}{\sin(2\alpha)}$, and this implies $\sin(2\alpha) = \frac{\sin(45°)}{\sqrt{2}} = \frac{1}{2}$, and since $\sin 30° = \frac{1}{2}$, $\alpha = 15°$.

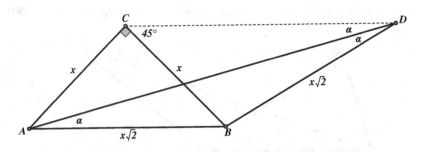

Figure 5.64

Solution 2: In Figure 5.65, we note that the altitude (y) in $\triangle ABC$ is half the length of its base AB. Then because $AB = BD$, we have the right triangle $\triangle BED$, with the hypotenuse BD twice the length of DE, which is a

---

[2]This law states that in a triangle $\triangle ABC$ with sides a, b, c, the following equation holds: $\frac{a}{\sin A} = \frac{b}{\sin B} = \frac{c}{\sin C}$.

property of a 30°–60°–90° triangle, where $\angle BDE = 60°$, so that $\angle BDC = \angle DBE = 30°$. Because of the equality of alternate-interior angles, we have $\angle ADC = \angle DAB = \angle ADB$. Therefore, $\angle ADB = \frac{1}{2}(30°) = 15°$.

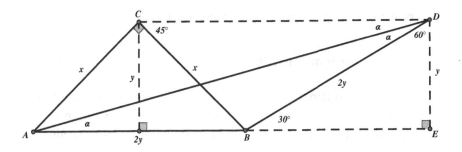

Figure 5.65

Determine the Fraction of the Rectangle that Is Shaded

In Figure 5.66, we show a rectangle with two tangent semicircles emitted from two consecutive vertices, C and D, of the rectangle. The common tangent to the two semicircles is BGH. Our challenge is to find the fraction of rectangle $ABCD$ which is shaded, namely, quadrilateral $ABHD$.

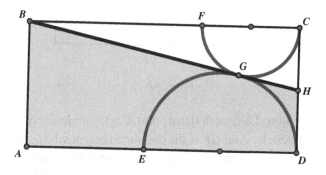

Figure 5.66

Solution: In Figure 5.66, we see that from the point H there are three tangential segments, HC, HG, and HD, which all have the same length. Therefore, H must be the midpoint of CD. The area of the rectangle whose length and width are BC and CH, is twice the area of triangle BCH. Hence, the area of $\triangle BCH = \frac{1}{4}$ area rectangle $ABCD$, whereupon it follows that the area of the shaded region is $\frac{3}{4}$ area rectangle $ABCD$.

Length of a Common Tangent

A semicircle and a quarter circle are inside a square, as shown in Figure 5.67. The radius of the semicircle is 5, and the radius of the quarter circle is 8. We need to determine the length of the common tangent whose endpoints G and F are on the opposite sides of the square.

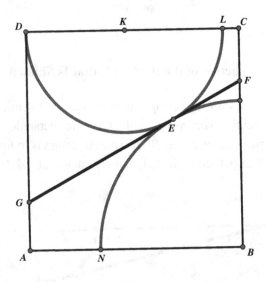

Figure 5.67

Solution: In Figure 5.68, we find that point K is the midpoint of the diameter DL of the semicircle. And GF is the common tangent whose length we are seeking. Since BE is perpendicular to GEF and KE is perpendicular to GEF, being radii to the points of tangency, we know that KEB is a straight line. Therefore, the length of BK is equal to the sum of the radii which is $5+8 = 13$. When we draw a line through C parallel to GF intersecting AD at point H, this creates parallelogram $GFCH$, which then gives us $CH = GF$. Furthermore, $\angle CKB + \angle KCH = 90°$ and $\angle CKB + \angle KBC = 90°$, therefore, $\angle KCH = \angle KBC$. This enables us to show that $\triangle CDH \cong \triangle BCK$. Thus, since $BK = CH$ and $CH = GF$, we then have $GF = BK = 13$, which satisfies our challenge.

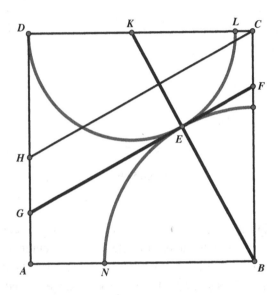

Figure 5.68

Two Circles and an Angle Bisector

In Figure 5.69, we see that circle c is internally tangent to a larger circle k at point P. We select any point Q on circle c. The tangent line to circle c at point Q intersects circle k at the points A and B. Quite unexpectedly, we note that PQ bisects $\angle APB$. Our challenge here is to verify this appearance.

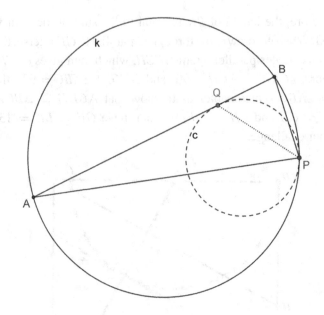

Figure 5.69

Solution: In Figure 5.70, we see that point E is the intersection of circle k and PQ. Also, RP is the diameter of circle c, whose center is point M, and PR extended intersects circle k at point G. Furthermore, the point D is the intersection point of AB and EM. Then we focus on the shaded quadrilateral $MRQD$, and in particular at its angles. If we can prove that $\angle MDQ = 90°$, then we would have completed the proof because EF is a diameter of k and this implies $AD = BD$ or in terms of arcs $AE = BE$, which then justifies the equality of the two inscribed angles, that is, $\angle APE = \angle BPE$.

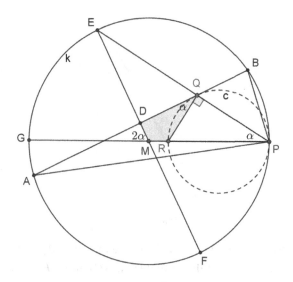

Figure 5.70

Fortunately, in quadrilateral $MRQD$, it is easy to see that $\angle MDQ = 90°$, which we will now pursue. Let α be $\angle RPQ$. Related to circle c we note that $\angle RQD = \angle RPQ = \alpha$ as they are both half the angle related to the arc QR. The exterior angle QRM of triangle QRP is equal to $90° + \alpha$. Since $\angle EMG$ is a central angle, we have $\angle EMG = 2\alpha$. Therefore, since $\angle EMP$ is supplementary to $\angle EMG$, we get $\angle EMP = 180° - 2\alpha$. Thus, the angle $\angle QDM = 360° - \alpha - (90° + \alpha) - (180° - 2\alpha) = 90°$. This then produces $AD = BD$ which, in turn, gives us arc AE equals arc BE, which finally — via the inscribed angle theorem — provides our desired result, namely, that $\angle APE = \angle BPE$.

Two Tangent Quarter Circles in a Rectangle

The bold line shown in Figure 5.71 is a common tangent of the two tangent quarter circles with centers at opposite vertices in a rectangle. We need to determine what part of the rectangle is the shaded area between the common tangent and the diagonal.

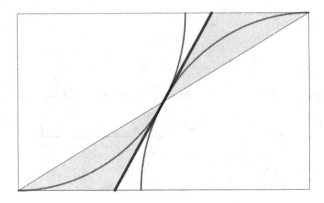

Figure 5.71

Solution: In Figure 5.72, let the width of the rectangle $ABCD$ with center M be 1. Then we can see that the triangles $\triangle AMD$ and $\triangle BMC$ are equilateral with side lengths 1, since the radii of the quarter circles are equal. Also, $BM = DM$ and $AM = CM$, as the diagonals bisect each other. This yields $a = \sqrt{3}$ for the length of the rectangle, which is comprised of two altitudes containing the point M of these two equilateral triangles. Since the radius is perpendicular to the tangent at the point of contact, we have right triangle $\triangle DME$ (with angles $30°$ and $60°$) so that we can compute $DE = \frac{2}{\sqrt{3}} = \frac{2}{3}\sqrt{3}$, and this means since $DC = a = \sqrt{3}$, then $EC = DC - DE = \sqrt{3} - \frac{2}{3}\sqrt{3} = \frac{1}{3}\sqrt{3} = \frac{1}{3}a$. Hence, the area of the shaded triangle $\triangle MCE$ is $\frac{1}{3}$ area $\triangle DMC = \frac{1}{3}\left(\frac{1}{4}\right)$ area $ABCD = \frac{1}{12}$ area $ABCD$, and since we have two such congruent triangles, the shaded area is $2\left(\frac{1}{12} \text{area } ABCD\right)$ or $\frac{1}{6}$ of the rectangle's area.

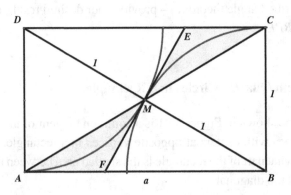

Figure 5.72

Alternative solution

In Figure 5.73, let the width of the rectangle $ABCD$ with center M be 1. Then we can see that the triangles $\triangle AMD$ and $\triangle BMC$ are equilateral with side lengths 1 since the radii of the quarter circles are equal. Also, $BM = DM$, and $AM = CM$ as the diagonals bisect each other. From that we can conclude that the triangles $\triangle AFM$ and $\triangle MCE$ are congruent isosceles triangles (ASA) with base angles $30°$, and we also have $\triangle FMG$ as a $30°$–$60°$–$90°$-triangle. Then reflecting the right triangle FMG in MF yields triangle $\triangle FMG'$ with the right angle $\angle FG'M$ at point G', which is the midpoint of AB (note: $\angle MFG = 60° = \angle BFM$). Therefore, area $\triangle AFM = \frac{2}{3}$ area $\triangle AG'M = \frac{2}{3} \cdot \frac{1}{8}$ area $ABCD = \frac{1}{12}$ area $ABCD$, and since we have two such congruent triangles, the shaded area is $2\left(\frac{1}{12}\right.$ area $\left. ABCD\right)$, or $\frac{1}{6}$ of the area of the rectangle.

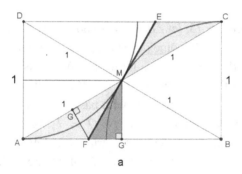

a

Figure 5.73

A Very Special Triangle

For triangle ABC, point D is the intersection point of the angle bisector of angle BAC with the side BC. The circumcenter, O, of the triangle $\triangle ABC$ coincides with the incenter, I, of the triangle $\triangle ADC$ (Figure 5.74). Our challenge is to find the measures of the angles of the triangle $\triangle ABC$.

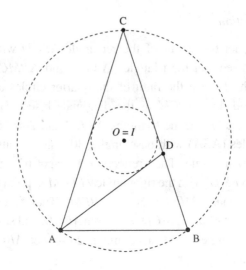

Figure 5.74

Solution: In Figure 5.75, let $\angle BAC = \alpha$ and the incenter I of $\triangle ADC$ and circumcenter O of $\triangle ABC$ coincide. Then, since AO bisects $\angle CAD$, we know $\angle BAO = \frac{3}{4}\alpha$, and $AO = CO$, we know that $\triangle AOC$ is isosceles with base angles of $\frac{\alpha}{4}$, and the same holds for $\triangle BOC$. We then can conclude that $\triangle ABO$ must be an isoseles triangle with base angles of $\frac{3}{4}\alpha$, where α is the measure of the vertex angle. (Note: At $O = I$, we have twice an angle of $\frac{\alpha}{2}$; these are exterior angles of the triangles $\triangle AOC$ and $\triangle BOC$; as a consequence, C, O/I, and the midpoint H of AB are collinear.) Since the angle sum in $\triangle ABO$ equals $180°$, we have $\alpha + (2 \times \frac{3}{4}\alpha) = 180°$ from which we get $\alpha = 72°$. Hence, the triangle $\triangle ABC$ must be a *golden triangle,*[3] since it is an isosceles triangle with base angles of $72°$ and a vertex angle of $36°$, which is a triangle that provides many further wonders!

---

[3]The golden triangle provides a plethora of amazing relationships which can be further investigated in *The Glorious Golden Ratio* by A.S. Posamentier and I. Lehmann (Prometheus Books, 2012).

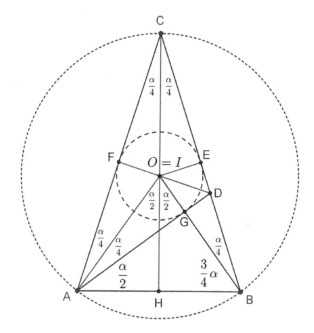

Figure 5.75

Chapter 6

For Experts: Stepwise Solutions to Challenging Problems

Problem solving is a time-honored and important issue in mathematics. Properly presented, problem solving should be an integral part of mathematics instruction. One can significantly benefit from being exposed to tasks for which the problem-solving technique may be new, insofar as it is not merely a repetition of previously exposed techniques. Of course, these tasks should be interesting and challenging, but on the other hand, they must not be too difficult because otherwise frustration will result. They should be enlightening!

The exposure to new problem-solving techniques should provide one with an understanding of what is meant by the *process of doing mathematics* and further appreciating the mathematical results by justifying them.

It is very important to find tasks that are "problems" in the sense that they promote creativity and spirit, motivate, are not limited to simple exercises, but are also not too tricky so as to provide one with a realistic chance at success. One can become self-confident and be further motivated by having successful experiences. We are considering motivating and interesting tasks that contain "real mathematics" that fit into the realm of problem solving and that provide a broad range of readers with an appetite for more mathematics.

Problem solving should not necessarily be treated as a mathematical topic as one would consider geometry, algebra, and calculus. Introducing problem solving should not be restricted to a series of techniques in which one is taught how to solve problems from a list of heuristic strategies or tell the students what the possible phases in a problem-solving process are. Ideally one should try to solve problems by oneself even though initially

277

it might be difficult to work independently. It might take quite some time for the average reader to have some success, but in the long run this time seems well spent towards experiencing the feeling what *doing mathematics* means.

We do not propose that all mathematics should be dedicated to problem solving since mathematics has many facts to be learned and many procedures to be practiced.

To fully appreciate problem-solving techniques, we present several challenging problems (listed as Problem 1, Problem 2, ...), each of which is partitioned into preparatory problems (listed as Problem 1a, 1b, 1c, ... for Problem 1 etc.) so that problem solvers can better appreciate the stepwise solution of the original problem.

In this final chapter, the problems are sometimes more advanced and of a somewhat higher mathematical level. Yet, we believe a motivated reader will be able to grasp the material quite well.

The Problems

Problem 1: Given a triangle ABC with incenter I, let D be the midpoint of the side AC and let G be the point of tangency of the incircle and the side AC. When DI is extended to meet GB, the point of intersection H is the midpoint of GB, as shown in Figure 6.1. Why is that true? (The analogous statement holds for the other two sides AB and BC as well.)

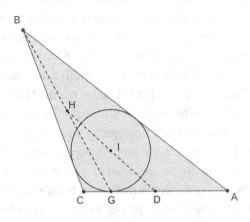

Figure 6.1

Before we employ a purely geometric solution without algebra and various calculations (which we will demonstrate a bit further on), we will present a solution that combines geometry and algebra.

The following series of problems will lead to a solution of Problem 1. The principle of considering a series of preparatory problems, sort of partitioning a problem, or a proof, into several parts, is often a good strategy for solving challenging problems.

Problem 1a: One can establish the formula for the inradius ρ of a right triangle $\rho = \frac{a+b-c}{2}$, where a and b are the lengths of the legs and the hypotenuse has length c, by an algebraic approach or just by inspecting appropriately Figure 6.2. It should be easy to explain by merely inspecting the diagram.

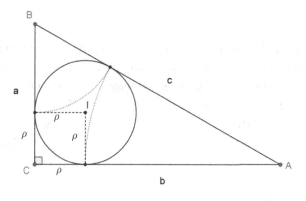

Figure 6.2

After drawing the radii of the inscribed circle from its center, I, to the points of tangency on lines BC and AC, we can easily see that since the tangent segments from each of the triangle vertices are equal, we get $a + b = c + 2\rho$, which is the equivalent of $\rho = \frac{a+b-c}{2}$.

Problem 1b: Analogously, one can establish the formula for the distance z of the vertex C from the points of tangency with the incircle $z = \frac{a+b-c}{2}$ by an algebraic approach or just by once again inspecting Figure 6.3. This is analogous to Problem 1a, the only difference is that in the general case this distance does not equal the inradius.

Problem 1c: Consider the situation shown in Problem 1, where triangle ABC has a right angle at vertex C, as seen in Figure 6.2.

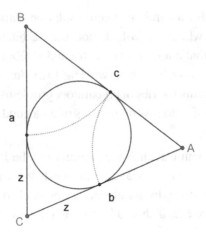

Figure 6.3

To prove that the extension of *DI* meets *GB* at its midpoint, we approach the problem in another fashion. As shown in Figure 6.4, we assume that *H* is the midpoint of *GB* and *J* denotes the intersection point of the horizontal line through *I* and the vertical line through *H*. We then prove that the triangles *JIH* and *GDI* are similar (that is to say that the slopes of *DI* and *IH* are equal). That means we have to show that $\frac{\frac{a}{2}-\rho}{\frac{\rho}{2}} = \frac{\rho}{\frac{b}{2}-\rho}$ holds.

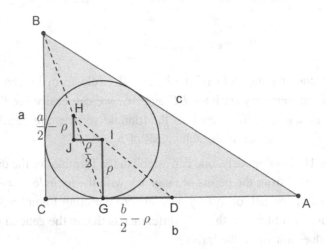

Figure 6.4

This can be rewritten as a quadratic equation in ρ:

$$\left(\frac{a}{2} - \rho\right) \cdot \left(\frac{b}{2} - \rho\right) - \frac{\rho^2}{2} = 0 \Leftrightarrow \frac{1}{2}\rho^2 - \left(\frac{a}{2} + \frac{b}{2}\right)\rho + \frac{ab}{4} = 0$$

$$\Leftrightarrow \rho^2 - (a+b)\rho + \frac{ab}{2} = 0 \qquad (*)$$

Using the quadratic formula, we get the solution: $\rho = \frac{a+b\pm\sqrt{a^2+b^2}}{2}$. Applying the Pythagorean theorem to triangle ABC, we have $\sqrt{a^2 + b^2} = c$ and because $\rho < \frac{a+b+c}{2}$ we can say that the above quadratic equation $(*)$ is equivalent to $\rho = \frac{a+b-c}{2}$. Our aim was to use this result also for the general case without further algebraic calculations: We wanted to show that the points H (as the midpoint of GB), I, and D stay collinear if the point B is moved horizontally in the height $a = h_b$ "above" b, but, unfortunately, we did not succeed. Now let us move towards the general case using a different approach with the next problem.

Problem 1d: Given a triangle with the sides a, b, c. Find a formula for the distance x from C to the foot F of the altitude h_b on the side b or on its extension, as shown in Figure 6.5.

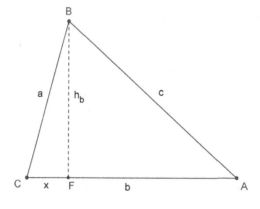

Figure 6.5

Let us first assume that F lies "to the right side of C," as shown in Figure 6.5. Applying the Pythagorean theorem twice to the two right triangles, BCF and ABF, yields: $(b - x)^2 = c^2 - \underbrace{(a^2 - x^2)}_{h_b^2}$, which then enables us to have $x = \frac{a^2+b^2-c^2}{2b}$. If F lies to the left of vertex C (Figure 6.6), then this value of x would be negative and we would get $|x| = \frac{c^2-a^2-b^2}{2b}$.

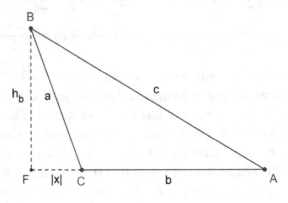

Figure 6.6

Now, we are ready to solve Problem 1 using Figure 6.7. Here we assume that F lies to the left of vertex C (the other case would work analogously).

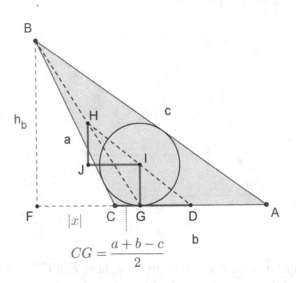

$$CG = \frac{a+b-c}{2}$$

Figure 6.7

With the same technique as in Problem 1c and with the well-known and easy-to-prove formula for the inradius ρ of a triangle, we have $\rho = \frac{A}{s} = \frac{2A}{2s} = \frac{b \cdot h_b}{a+b+c}$, where A denotes the area and s denotes the semi-perimeter. Now we need to show the following (applying similar triangles or using the concept of equal slopes):

$$\frac{\frac{h_b}{2} - \rho}{\frac{|x| + \frac{a+b-c}{2}}{2}} = \frac{\rho}{\frac{b}{2} - \frac{a+b-c}{2}} \qquad (**)$$

At first glance, this seems rather complicated, but it is not; by cross multiplying we get a linear equation for $|x|$:

$$\frac{\left(\frac{h_b}{2} - \rho\right)\left(\frac{b}{2} - \frac{a+b-c}{2}\right)}{\frac{\rho}{2}} = |x| + \frac{a+b-c}{2} \Longleftrightarrow$$

$$|x| = \left(\frac{h_b}{2\rho} - 1\right)(c-a) - \frac{a+b-c}{2}$$

Using $\rho = \frac{b \cdot h_b}{a+b+c}$, we get after simplification $|x| = \frac{c^2 - a^2 - b^2}{2b}$, which we know from Problem 1d. If one accepts that x is negative as in case of F lying on the left side of C, there is no need for distinguishing the two cases. In the equation (**), one would have to write $-x$ instead of $|x|$.

Now consider a completely different, yet more elegant, synthetic solution to Problem 1, where we will use some preparatory problems.

Problem 1e: Here we will prove that for triangle ABC the two distances of the tangency points of the incircle and the excircle from the two vertices A and C are equal, which we see in Figure 6.8, where we will prove that $CG = AG_1$.

Let G and G_1 be the two tangency points (on the side CA) of the incircle and the excircle, respectively, as shown in Figure 6.8. Then since the tangent segments of a circle from an external point are equal, we have $BR = BS$ and $BP = BQ$, and furthermore by subtraction $PR = QS(*)$.

Furthermore, $PC = CG, PR = PC + CR = PC + CG + GG_1 = 2CG + GG_1$, and similarly, we have $QS = QA + AS = GG_1 + G_1A + AS = 2AG_1 + GG_1$. Together with (*) we can conclude that $CG = AG_1$.

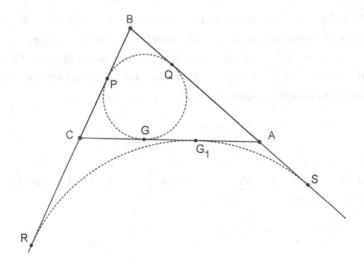

Figure 6.8

Problem 1f: Let G be the point of tangency of the incircle on the side CA; we will consider G as the "south pole" of the incircle and denote its "north pole" as G'. Our challenge is to prove that in this situation the points B, G', G_1 are collinear (Figure 6.9), where G_1 denotes the point of tangency of the excircle.

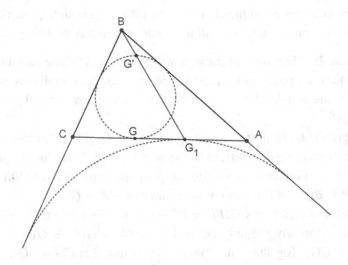

Figure 6.9

We will use triangle similarity, and let G_1 be the intersection point of BG' with AC. We then have to prove that G_1 is the point of tangency of AC with the excircle. From G_1, we draw a perpendicular to AC and intersect it with the angle bisector BM_1 of the angle at B, which yields M_1 (Figure 6.10), and we must show that M_1 is the center of the excircle. The triangles $\triangle BMG'$ and $\triangle BM_1G_1$ are similar, where M denotes the incenter of $\triangle ABC$. We define $k = \frac{BM_1}{BM}$. Then we have $G_1M_1 = k \cdot \underbrace{G'M}_{\rho}$ (here ρ

represents the inradius, $G'M = SM$, of triangle ABC), and also the triangles BMS and BM_1S_1 are similar, as well are triangles BMR and BM_1R_1. Hence, we have also $S_1M_1 = k \cdot \underbrace{SM}_{\rho}$ and $R_1M_1 = k \cdot \underbrace{RM}_{\rho}$. This means that

$G_1M_1 = S_1M_1 = R_1M_1 (= k \cdot \rho)$, and because we have right angles at R_1, G_1, S_1, we, therefore, know that M_1 is the center of the excircle.

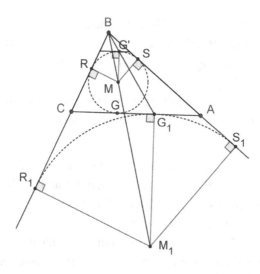

Figure 6.10

Now we are prepared to solve Problem 1 in another way, synthetically, without algebra: see Figure 6.11.

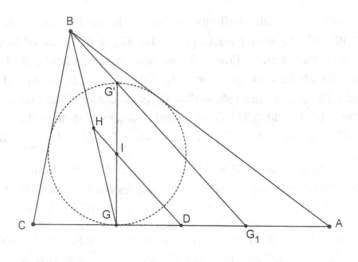

Figure 6.11

Since D is the midpoint of GG_1 (Problem 1e) and I is the midpoint of GG', we see that $DI||G_1G'$. And because of the collinearity of B, G', G_1 (Problem 1f), considering triangle BGG_1, the extension of DI meets BG at its midpoint (H).

Another example for the principle of the stepwise solution is the following. Here algebraic calculations do not play a major role as this problem is largely geometric.

Problem 2: Let $ABCD$ be a cyclic quadrilateral and S, T, U, V the incenters of the four triangles, $\triangle ABD$, $\triangle ABC$, $\triangle BCD$, and $\triangle ADC$, which result from the two diagonals of quadrilateral $ABCD$. Prove that $STUV$ is a rectangle, which is shown in Figure 6.12.

This problem has been used for training students for mathematics Olympiad competitions and is probably a bit of a challenge for a normal geometry class. But again, one could think of solving a series of subsequent problems — better accessible for students and easier to solve — which will lead to a solution of Problem 2.

Here is a series of problems (2a, 2b, 2c) which prepare us for the solution of Problem 2.

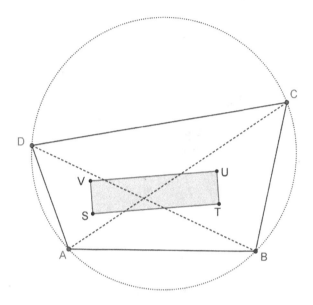

Figure 6.12

Problem 2a: Cloverleaf figure in a circle

On a circle c, shown in Figure 6.13, where we select the circular arcs AB with center M_1 and BC with center M_2 are drawn which intersect at a point S in the interior of circle c. Let M_3 be the midpoint of the arc AC on circle c, which does not contain M_1, M_2. Then we can also consider the arc CA with center M_3 running in the interior of circle c, as shown in Figure 6.13. The result is a figure which seems to consist of three clover leaves (in other words, three circular arcs intersecting each other at S). All three clover leaves with the vertices A and S, B and S, and C and S consist of two intersecting circular arcs which determine equal angles at both vertices due to symmetry. The angles between the corresponding *tangents* are shown as dotted lines. These angles we will call *clover angles* of the clover leaves.

We shall prove the following statements:

(1) The straight lines M_1C and M_2A pass through S.
(2) The straight line M_3B and the arc CA with center M_3 pass through S.

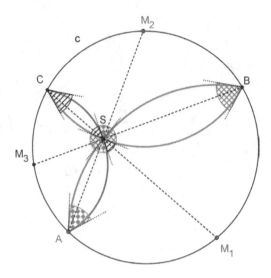

Figure 6.13

(3) S is the incenter of the triangle ABC.
(4) The sum of the clover angles at the vertices of the clover leaves[1]
     A, B, C is $180°$.

Following are the proofs of the four above statements:

(1) First, we will prove that M_1C passes through S. Since M_1 is the midpoint of $\overparen{ABC}$ and M_2 is the midpoint of $\overparen{BSC}$, we have $M_1S = M_1B$ and $M_2S = M_2B$. Hence, the quadrilateral SM_1BM_2 is a kite, and thus we have $\angle BM_1M_2 = \angle SM_1M_2$. We also know that $\angle BM_1M_2 = \angle CM_1M_2$ (from the equal arcs $\overparen{BM_2}$ and $\overparen{M_2C}$) and this yields $\angle SM_1M_2 = \angle CM_1M_2$ which then determines the collinearity of M_1, S, C, as shown in Figure 6.14. Analogously, one can prove the collinearity of M_2, S, A.

(2) Since M_2 is the midpoint of $\overparen{BC}$, we know that $\overparen{M_2B} = \overparen{M_2C}$ and the angles inscribed in these equal arcs are are equal, so that, $\angle CAM_2 = \angle BAM_2$, in other words AM_2 is the angle bisector of $\angle CAB$ containing S (note the collinearities of 1). Analogously, we see that CM_1 is the angle bisector of $\angle ACB$ containing S. Therefore,

---

[1]This item is not necessary for Problem 2; it is merely offered as further enrichment.

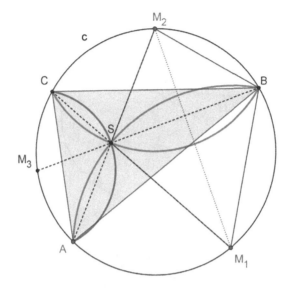

Figure 6.14

S must be the incenter of $\triangle ABC$ and it must also lie on the angle bisector of angle B, which proves both the collinearity of M_3, S, B and that establishes that S is the incenter of triangle ABC.

(3) Point S was just shown to be the incenter of triangle ABC in the above point.

(4) Due to symmetry, each of the three leaves has equal clover angles (more precisely angles between the respective tangents) at both vertices. At point S, we find that the vertical angles of the intersecting tangents produce equal angles as shown in Figure 6.13. Furthermore, at point S, these angles have a sum of $360°$, that is, the three clover angles and the three equal corresponding vertical angles between the leaves. Thus, the three clover angles at S have a sum of $180°$. Since the three clover angles at the respective other vertices (A, B, C) are the same, their sum is also $180°$.

Problem 2b: We know that two parallel lines cut off equal arcs on a circle as shown in Figure 6.15. The challenge here is to prove that two chords e, f of a circle are perpendicular if, and only if, the sum of opposite arc lengths is equal: $e \perp f \Leftrightarrow a + c = b + d$ (see Figure 6.16).

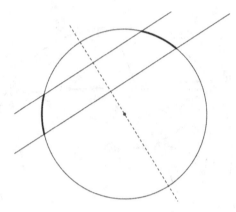

Figure 6.15

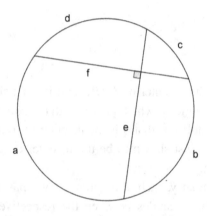

Figure 6.16

We begin with the chords that are two perpendicular diameters and shift first one diameter to a parallel chord and then the other to another parallel chord. In the first situation of two perpendicular diameters, all four arc lengths are equal (a quarter of the perimeter), hence, we have $a+c = b+d$. If one diameter is moved to a parallel chord (Figure 6.17), a, b become bigger by the same amount as c, d become smaller. Thus $a + c = b + d$ still holds. Using the same idea, we get that after the movement of the second diameter to a parallel chord $a + c = b + d$ also holds true (Figure 6.18).

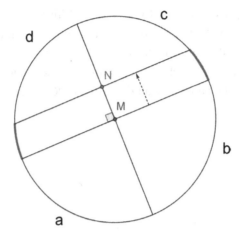

Figure 6.17

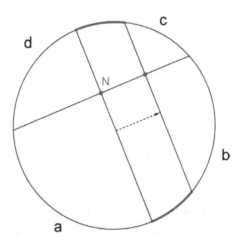

Figure 6.18

Shifting chords in a parallel manner, as shown in Figure 6.18, does not affect the sums $a + c$ and $b + d$. Conversely, if $a + c = b + d$ holds, we can move (shift) both chords to the center of the circle and here we see that the corresponding diameters must be perpendicular.

Problem 2c: Let $ABCD$ be a cyclic quadrilateral with circumcircle c. These four vertex points divide circle c into four arcs with midpoints

M_1, M_2, M_3, M_4, as shown in Figure 6.19. With these points as centers, we draw four arcs in the interior of circle c that have the intersection points S, T, U, V. We shall prove that these points S, T, U, V are the incenters of the four large triangles which are determined by the diagonals of the cyclic quadrilateral $ABCD$, namely, triangles ABC, BDC, ABD, and ADC.

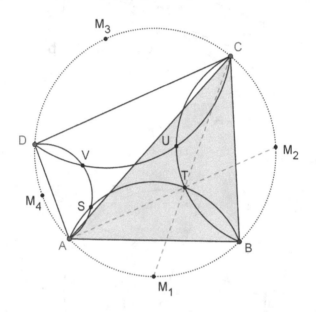

Figure 6.19

As we can see in Figure 6.19, T is the intersection point in the circle interior of the arc AB centered at M_1 and the arc BC centered at M_2. From Problem 2a, we know that T is the point of intersection of the angle bisectors of triangle ABC, which is then the incenter of triangle ABC. The same holds true for the other three points S, U, V.

Now we are prepared to solve Problem 2 referring to Figure 6.20.

First, we note that $M_1 M_3 \perp M_2 M_4$ because the sum of two opposite intercepted arcs is in both cases $a + b + c + d = 180°$ (from Problem 2b). Moreover, we have two equal arcs $\widehat{DM_3} = \widehat{CM_3}$ and due to the inscribed angle relationship we can conclude that $M_1 M_3$ is the angle bisector of $\angle CM_1 D$. Furthermore, T is on $M_1 C$ and S is on $M_1 D$, hence, we can also state that $M_1 M_3$ is the angle bisector of $\angle T M_1 S$. The triangle $T M_1 S$ is isosceles, and

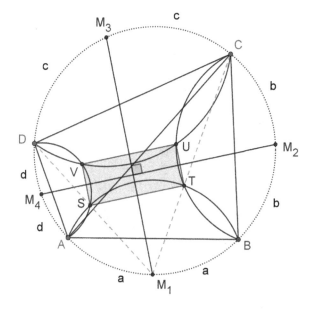

Figure 6.20

thus, we have $M_1 M_3 \perp ST$. We can repeat this procedure with the other sides of the quadrilateral $STUV$, which then completes the proof.

In the following, we introduce several other such challenging problems, which are treated in a similar way: Formulating a series of problems whose solutions cleverly lead to the solution of the initial problem. These problems should be interesting and provide some interesting insights and should be easier to solve by considering the stepwise approach we used previously.

Problem 3: The circle inscribed in quadrilateral $ABCD$ is tangent to its sides at points X, Y, Z, W, as shown in Figure 6.21. We then find that the lines XZ and WY pass through the point M, which is the point of intersection of diagonals AC and BD.

Problem 3a: The circle inscribed in quadrilateral $ABCD$ is tangent to its sides at points X, Y, Z, W, as shown in Figure 6.22. Then the line XZ divides the diagonal AC in the ratio $\frac{AN}{CN} = \frac{AX}{CZ}$.

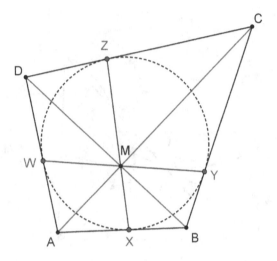

Figure 6.21

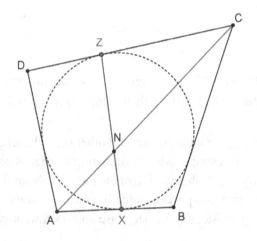

Figure 6.22

Let N be the point of intersection of AC and XZ. Since an angle formed by a tangent and a chord is one-half the intercepted arc, we have $\angle AXN = \frac{1}{2}$ arc $XZ = \angle NZD$. Furthermore, supplementary angles have equal sines, hence

$$\sin \angle AXN = \sin \angle NZD = \sin \angle NZC \qquad (1)$$

We then can conclude the following:

$$2 \cdot \text{area}(\triangle AXN) = (AN)(XN)\sin \angle ANX = (AX)(XN)\sin \angle AXN$$

For $\triangle NCZ$ we have analogously

$$2 \cdot \text{area}(\triangle NCZ) = (CN)(ZN)\sin \angle CNZ = (ZN)(ZC)\sin \angle NZC$$

Taking the quotient of the last two lines yields, because of (1) and $\angle ANX = \angle CNZ$ (vertical angles), $\frac{(AN)(XN)}{(CN)(ZN)} = \frac{(AX)(XN)}{(ZN)(ZC)}$. And from that we conclude immediately $\frac{AN}{CN} = \frac{AX}{CZ}$, as claimed.

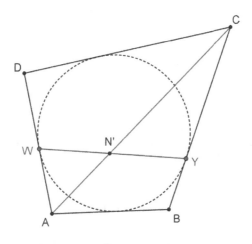

Figure 6.23

Problem 3b: Use the solution of Problem 3a to prove Problem 3.

Using an analogous argument, now assume that AC and WY intersect at some point, say N', as shown in Figure 6.23. Considering the triangles $AN'W$ and $N'YC$, we can prove that WY divides AC in the ratio $\frac{AM}{CM} = \frac{AW}{CY}$. However, $AW = AX$ and $CY = CZ$ because tangent segments from the same point to a circle are equal. In other words, the lines XZ and WY divide AC in the *same* ratio, and that means $N = N'$. We see that the diagonal AC passes through the point of intersection of XZ and WY. And if we use the same procedure with the diagonal BD instead of diagonal AC, we can see that the same holds for the diagonal BD. This completes the proof of Problem 3.

Problem 4: Let c_1 and c_2 be two circles intersecting at points A and B and a straight line through A (other than AB) is drawn, intersecting the two circles at points M and N, as shown in Figure 6.24. Furthermore, let K be the midpoint of MN, P the intersection point of the angle bisector of $\angle MAB$ with circle c_1, and R the intersection point of the angle bisector of $\angle BAN$ with circle c_2. We need to prove that $\angle PKR = 90°$.

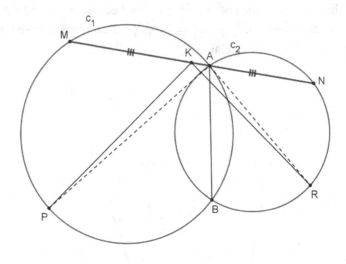

Figure 6.24

This problem seems not to be too complicated, but despite its elementary appearance, the problem proves to be deceptively hard and may resist several approaches.[2] In the end, there are many different solutions. We offer the shortest solution using another problem, but for some problem solvers, it may be difficult to realize that these steps have something to do with the original problem.

Problem 4a: Let $\triangle ABC$ be a right triangle, and congruent right triangles $\triangle CDE$ and $\triangle BFG$ similar to $\triangle ABC$, as shown in Figure 6.25. We have $\triangle CDE$ and $\triangle BFG$, which are placed so that their corresponding sides are

---

[2]For details, see De Villiers, M., Humenberger, H. (2021). "Ghosts of a problem past." *At Right Angles*, Issue 9, 105–111.

parallel. Then we will prove that triangle AFE is also a right triangle and similar to $\triangle ABC$.

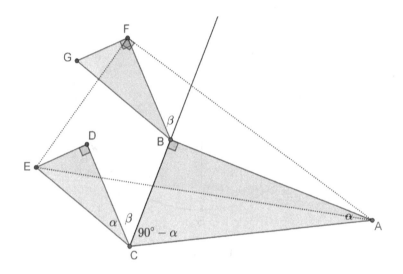

Figure 6.25

For a proof of Problem 4a, as shown in Figure 6.25, let $\angle CAB = \alpha$ and observe that $\triangle AFB$ and $\triangle AEC$ are similar, as they have equal angles at B and C, namely, $90° + \beta$, and then the ratio of their corresponding sides is equal: $\frac{AB}{AC} = k = \frac{BF}{CE}$. Therefore, also $\frac{AF}{AE} = k$ and $\angle EAF = \alpha$, and thus, the claimed similarity is proved.

Problem 4b: In a triangle, the perpendicular bisector of a side and the angle bisector of the opposite angle intersect at a point on the circumcircle, as shown in Figure 6.26 (since S is a sort of *south pole* in the figure, one could call this result *south pole theorem*).

Let S be the intersection point of the circumcircle with the perpendicular bisector of AB, then we have equal arcs $SA = SB$; with equal arcs, the corresponding inscribed angles must be equal, that means $\angle ACS = \angle BCS$, hence, CS must be the angle bisector.

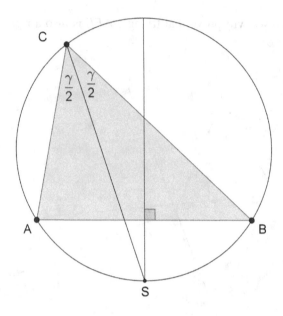

Figure 6.26

Problem 4c: We will use the solution of Problems 4a and 4b to solve Problem 4.

The plan here is to construct triangle *MBN* and the midpoints *H* and *J* of the line segments *MB* and *BN*, respectively. We then focus on the triangles *BRJ* and *PBH*. One just has to use the result of Problem 4a once (see Figure 6.27). In retrospect, things often seem to be very easy, but to *find* these simple relations is often not trivial, as Problem 4 is definitely a really hard problem from the perspective of an initial solver.

Now, for the proof, consider Figure 6.27, where *H* is the midpoint of *BM* and *J* is the midpoint of *BN*, and let $\alpha = \angle BRJ$ in the right triangle *BRJ*. (Recall that $\triangle BRJ$ and $\triangle PBH$ are right triangles because of Problem 4b.) Then *KJ* is equal and parallel to *BH*. Because *BRNA* is a cyclic quadrilateral, we have $\angle BAN = 180° - 2\alpha$, and from this we get (from the cyclic quadrilateral *PBAM*) $\angle MPB = 180° - 2\alpha$, and this yields $\angle HPB = 90° - \alpha$. Then we draw *QJ* parallel and equal to *PB*, and *QK* parallel and equal to *PH*, in order to get the third right triangle ($\triangle QJK$, legs *KJ* and *QK*), all of them with angles $90°$, α, $90° - \alpha$. And by using Problem 4a, it follows immediately that $\angle PKR = 90°$.

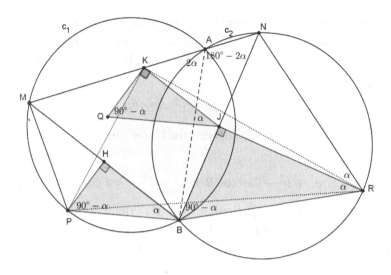

Figure 6.27

Problem 5: A *bicentric quadrilateral* is one which has both an incircle and a circumcircle. Show that the area of a bicentric quadrilateral with sides a, b, c, d is given by the formula $F = \sqrt{abcd}$.

Remark: This formula is not only valid for bicentric quadrilaterals, but also rectangles have this area formula.

Problem 5a: We will show that for a bicentric quadrilateral, as shown in Figure 6.28, the following equation holds: $(ab + cd) \cos B = ab - cd$.

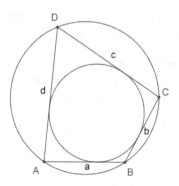

Figure 6.28

From the well-known property of tangential quadrilaterals, $a + c = b + d$, (note that tangential segments from an exterior point to a circle are equal) we immediately get $a - b = d - c$, and if we square both sides of this equation, we get

$$a^2 - 2ab + b^2 = c^2 - 2cd + d^2 \qquad (*)$$

Drawing the diagonal AC and applying the *law of cosines*[3] to triangles ABC and ACD yield

$$a^2 + b^2 - 2ab \cos B = c^2 + d^2 - 2cd \cos D \qquad (**)$$

The subtraction of the equations, $(**) - (*)$, yields $2ab(1 - \cos B) = 2cd(1 - \cos D)$, and because $\cos D = -\cos B$, we have $ab(1 - \cos B) = cd(1 + \cos B)$ and this is equivalent to $ab - cd = (ab + cd) \cos B$, which is what we sought to prove.

Problem 5b: Show that for a bicentric quadrilateral, the following equation holds: $2F = (ab + cd) \sin B$ (where F denotes the area of the quadrilateral)

This is a consequence of the well-known triangle area formulas: $\frac{ab}{2} \sin B = F = \frac{cd}{2} \sin D$ and $\sin B = \sin D$. (Note: The measures of the angles at B and D have a sum of $180°$ because $ABCD$ is cyclic.)

Problem 5c: Use the solution of Problems 5a and 5b to solve Problem 5.

From Problem 5b, we get $4F^2 = (ab + cd)^2 \underbrace{(1 - \cos^2 B)}_{\sin^2 B}$ and this yields $4F^2 = (ab + cd)^2 - (ab + cd)^2 \cos^2 B$; now, with the help of Problem 5a, we substitute $(ab + cd)^2 \cos^2 B$ by $(ab - cd)^2$ and get $4F^2 = (ab + cd)^2 - (ab - cd)^2 = 4abcd$. Finally, dividing by 4 and taking the square root gives us the desired result: $F = \sqrt{abcd}$.

Problem 6: Take an arbitrary initial natural number N_0. We state the following algorithm: Every digit of this natural number is multiplied by 111 and then the sum of all these products is taken.

Example: $N_0 = 4367$, then $4 \cdot 111 + 3 \cdot 111 + 6 \cdot 111 + 7 \cdot 111 = 444 + 333 + 666 + 777 = 2220 = N_1$

---

[3]The law of cosines states that in triangle $\triangle ABC$ with sides a, b, c, the following equation holds: $c^2 = a^2 + b^2 - 2ab \cos C$.

Then the same occurs to this number 2220:

$$N_1 = 2220, \text{ then } 2 \cdot 111 + 2 \cdot 111 + 2 \cdot 111 + 0 \cdot 111 = 222 + 222 + 222 = 666 = N_2$$

And then to this number the same procedure is applied:

$$N_2 = 666, \text{ then } 6 \cdot 111 + 6 \cdot 111 + 6 \cdot 111 = 666 + 666 + 666 = 1998 = N_3$$

And another time:

$$N_3 = 1998, \text{ then } 1 \cdot 111 + 9 \cdot 111 + 9 \cdot 111 + 8 \cdot 111$$
$$= 111 + 999 + 999 + 888 = 2997 = N_4$$

Repeating this procedure with 2997, will yield 2997, so that the following pattern of results evolves as:

$$4367 \to 2220 \to 666 \to 1998 \to 2997 \to 2997 \to \cdots$$

We must prove that all positive integers converge to 2997 by applying this algorithm (not necessarily within four steps).

Problem 6a: First we prove the following: If one takes an arbitrary positive integer N_0 with at most three digits, the iteration of the algorithm will always end up at the number 2997 after at most four steps.

If the above-mentioned algorithm is considered as a function f, one can say $N_1 = f(N_0)$.

We can write f in another way: $f(N) = 111 \cdot DS(N)$, where DS denotes the digit sum. Because of the given start condition, we know for all three-digit numbers $DS(N_0) \leq 27$.

From $N_1 = 111 \cdot \underbrace{DS(N_0)}_{\leq 27}$, we know $N_1 \leq 111 \cdot 27 = 2997$. What can be said about $DS(N_1)$?

There are only a few numbers ≤ 2997 with digit sum >27: 1999, 2899, and 2989. But these numbers cannot serve as N_1 because they are not divisible by 111 (not even divisible by 3). Therefore, we can say $DS(N_0) \leq 27 \Rightarrow DS(N_1) \leq 27$, analogously, in the further steps: All N_i stay ≤ 2997 and all $DS(N_i)$ stay ≤ 27.

On the other hand, the following statements hold:

- 3 divides N_1 (independent of N_0) because $N_1 = 111 \cdot DS(N_0)$. Since the first factor of this product, 111, is divisible by 3, also the product N_1 itself must be divisible by 3.
- 3 divides $N_i \Rightarrow$ 9 divides N_{i+1} because $N_{i+1} = 111 \cdot DS(N_i)$ and we know that the first factor 111 is divisible by 3 and the second factor $DS(N_i)$ is also divisible by 3 due to the precondition that 3 divides N_i.
- 9 divides $N_i \Rightarrow$ 27 divides N_{i+1} because $N_{i+1} = 111 \cdot DS(N_i)$ and we know that the first factor 111 is divisible by 3 and the second factor $DS(N_i)$ is divisible by 9 due to the precondition that 9 divides N_i (note that N_i is divisible by 9 if, and only if, its digit sum $DS(N_i)$ is divisible by 9).

That means that if N_0 is not a multiple of 3, every step of the transitions $N_0 \rightarrow N_1 \rightarrow N_2 \rightarrow N_3$ produces one more factor of 3 in the number. Therefore, $N_3 = 111 \cdot DS(N_2)$ is surely divisible by 27 due to the second factor $DS(N_2)$ and by 111 due to the first factor 111. This means N_3 is surely divisible by 999, so there are only three possibilities for N_3 namely, $N_3 = 999, 1998, 2997$, and in the first two cases, the *next* step produces 2997.

If N_0 itself is divisible by 3 or even by 9, one ends up at 2997 even earlier.

Problem 6b: Prove that the same holds for arbitrary N_0 with at most 10,010,010 digits.

Let N_0 have at most 10,010,010 digits, then we have $DS(N_0) \leq 90,090,090$ and we can conclude $N_1 = 111 \cdot DS(N_0) \leq 111 \cdot 90,090,090 = 9,999,999,990$. Thus, N_1 has at most 10 digits with $DS(N_1) < 90$. Analogously, we can conclude $N_2 = 111 \cdot DS(N_1) < 111 \cdot 90 = 9,990$, thus, N_2 has at most four digits, that is, $DS(N_2) < 36$, and we know that 9 divides N_2, which is equivalent to 9 divides $DS(N_2)$, that means $DS(N_2) = 9, 18, 27$. Thus, there are only three possibilities for $N_3 : 999, 1998,$ and 2997 and N_4 is certainly 2997.

We can use the solution of Problems 6a and especially 6b for solving Problem 6, which is now provided.

Proof for Problem 6: All integers with at least five digits become smaller using f (most of them significantly smaller), thus, they eventually end up in the above-mentioned region, that is, numbers with at most 10,010,010 digits, as per Problem 6b:

Let N be a positive integer with n digits $(n \geq 5)$: $10^{n-1} \leq N < 10^n$, $DS(N) \leq 9 \cdot n$.

Therefore, $\frac{f(N)}{N} < \frac{10^3 \cdot n}{N} \leq \frac{10^3 \cdot n}{10^{n-1}} = \frac{n}{10^{n-4}}$ and that is <1 for all $n \geq 5$, thus, $f(N) < N$ as claimed.

We leave to the reader another challenging problem, and one worth trying, that is, to find a natural number, which needs five or six steps to arrive at 2,997, and probably a much more difficult problem to find the *smallest* such numbers.

Problem 7: In Figure 6.29, we are given a non-cyclic convex quadrilateral $ABCD$. Then the perpendicular bisectors of the sides are constructed. The two perpendicular bisectors adjacent to the vertex A (namely, $A'B'$ perpendicular to AB, and $A'D'$ perpendicular to AD) intersect at point A'. Similarly, for the other vertices generating quadrilateral $A'B'C'D'$, which we could call "Pb[4]-quadrilateral" to $ABCD$. In the same way $A''B''C''D''$

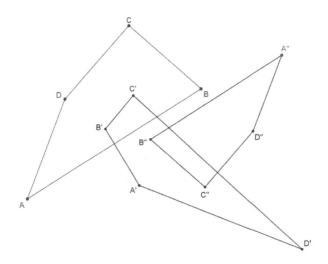

Figure 6.29

---

[4]From "perpendicular bisectors."

results from the intersections of the perpendicular bisectors of quadrilateral $A'B'C'D'$. We need to prove that $A''B''C''D''$ is similar to $ABCD$ with parallel corresponding sides. (Note: In Figure 6.29, perpendicular bisectors are not extended to their perpendicular lines here in order to keep the figure as clear and simple as possible.)

Problem 7a: We need to prove that in the context of Problem 7, the exterior angles of $A'B'C'D'$ are equal to the interior angles of $ABCD$, as shown in Figure 6.30.

Because of the right angles at the midpoints of the sides AB and BC, the angle at B' (interior angle of $A'B'C'D'$) is supplementary to the angle at B (interior angle of $ABCD$, shown in Figure 6.30). Since the sum of interior angles in every convex quadrilateral is $360°$. Angles X and Y are right angles of quadrilateral $BXB'Y$, the remaining two angles $B'YB$ and XBY are supplementary. Similarly, the angles at C and C' of quadrilateral $CZC'X$ are supplementary, as are the pairs of angles at D and D' as well as at A and A'. Therefore, the exterior angles of $A'B'C'D'$ are equal to the interior angles of $ABCD$, as claimed.

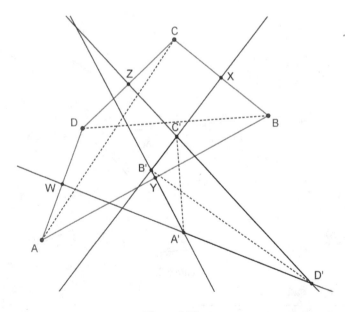

Figure 6.30

Problem 7b: Prove that in the context of Problem 7, the diagonals of $ABCD$ have the same angles as the diagonals of $A'B'C'D'$.

We will show that the diagonals of $A'B'C'D'$ are perpendicular to the diagonals of $ABCD$ as follows: The distances from C' to C and from C' to D are equal, and the point C' is also equidistant from points B and C, since the points on the perpendicular bisector of a line segment are equidistant from the endpoints. Thus, the distances of point C' to points B and D must be equal as well. Analogously, one can argue that A' is equidistant from points B and D, and thus $A'C' \perp BD$ because A' and C' are points on the perpendicular bisector of BD. In the same way, one can show that $B'D' \perp AC$, which yields the claimed perpendicularity of the diagonals of $A'B'C'D'$ to the diagonals of $ABCD$. And this, in turn, means that the diagonals of $ABCD$ have the same angles as the diagonals of $A'B'C'D'$, as originally claimed.

Problem 7c: Use the solution of Problems 7a and 7b for solving Problem 7.

From Problem 7b we know that corresponding diagonals and sides (of the quadrilaterals $ABCD$ and $A'B'C'D'$) are perpendicular, and this, in turn, means that corresponding diagonals and sides in the quadrilaterals $ABCD$ and $A''B''C''D''$ are parallel. Hence, corresponding triangles consisting of two sides and a diagonal of the quadrilaterals $ABCD$ and $A''B''C''D''$ are similar to each other with the same factor of similitude because the two triangles with the diagonal AC ($\triangle ABC$ and $\triangle ACD$) have this diagonal as a common side. Thus, the quadrilaterals $ABCD$ and $A''B''C''D''$ are similar, as claimed.

Problem 8: In cyclic quadrilateral $ABCD$, where the opposite sides are not parallel, let P and Q be the intersection points of the extensions of pairs of opposite sides. Then angle bisectors at P and Q are constructed and intersect the sides of $ABCD$ at the intersection points E, F, G, and H, as can be seen in Figure 6.31. We shall prove that $AC\|FG\|EH$ and $BD\|EF\|GH$.

Problem 8a: Prove that in the context of Problem 8, the two angle bisectors at points P and Q are perpendicular, as shown in Figure 6.31 (we had already discussed this problem in the section "Perpendicular Bisectors of Extended Sides of a Cyclic Quadrilateral" on page 230).

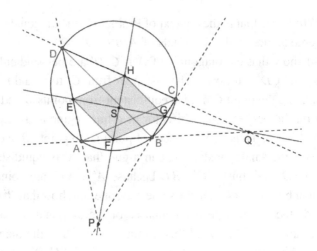

Figure 6.31

Let S be the intersection point of the two angle bisectors and A, B, C, D denote not only the vertices but also the corresponding interior angles of the quadrilateral $ABCD$. We begin by considering the concave quadrilateral $PBQS$, seen in Figure 6.31, and triangle DPC, where we have $\angle DPB = 180° - (C + D)$. Then, $\frac{1}{2}\angle DPB = \angle SPB = 90° - \frac{C+D}{2}$. Analogously, we have $\angle SQB = 90° - \frac{A+D}{2}$. Then for quadrilateral $PBQS$, $\angle PBQ = 360° - B$. We then have

$$\angle PSQ = 360° - (360° - B) - \left(90° - \frac{C+D}{2}\right) - \left(90° - \frac{A+D}{2}\right)$$

$$= \underbrace{B+D}_{180°} + \overbrace{\frac{A+C}{2}}^{180°} - 180° = 90°.$$

Note that because $ABCD$ is cyclic, $A + C = 180° = B + D$.

Problem 8b: Prove that in the context of Problem 8, the quadrilateral $EFGH$ is a rhombus.

The triangles PSE and PSG are congruent (ASA), hence, $ES = SG$. Analogously, the triangles QSF and QSH are congruent, then $FS = SH$. That means that in the quadrilateral $EFGH$, the diagonals are perpendicular and they bisect each other, thus, it is a rhombus.

Problem 8c: Use the solution of Problems 8a and 8b for solving Problem 8. For $AC||FG$ we have to show $\frac{a}{b} = \frac{d}{c}$ (Figure 6.32).

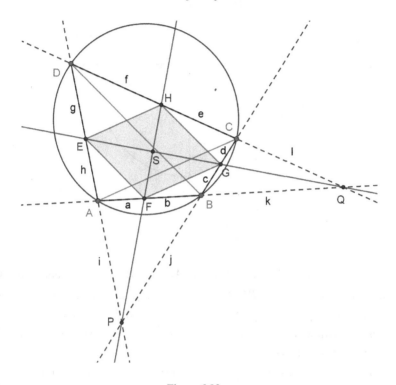

Figure 6.32

Recall that an angle bisector of a triangle divides the opposite side in the ratio of the adjacent sides, therefore, for the two angle bisectors of triangles ABP and BCQ, we have $\frac{a}{b} = \frac{i}{j}$ and $\frac{d}{c} = \frac{l}{k}$, respectively. For our goal $\frac{a}{b} = \frac{d}{c}$, we have to show $\frac{i}{j} = \frac{l}{k}$. This relation holds on the one hand because of the law of sines[5], and on the other hand due to $\sin(180° - \varphi) = \sin(\varphi)$ which is shown in Figure 6.33.

---

[5]The law of sines says that in triangle ABC with sides a, b, c, the following equation holds: $\frac{a}{\sin A} = \frac{b}{\sin B} = \frac{c}{\sin C}$.

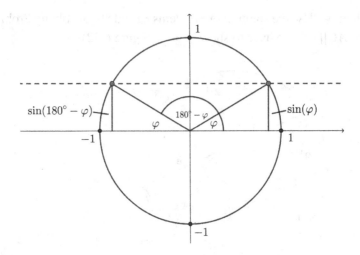

Figure 6.33

Applying the law of sines to triangle APB yields $\frac{i}{j} = \frac{\sin B}{\sin A}$, and applying the law of sines to triangle BCQ, we get $\frac{l}{k} = \frac{\sin B}{\sin C}$, and since $\sin A = \sin C$, both ratios are equal. This proves $\frac{a}{b} = \frac{d}{c}$ so that in the triangle ABC we get the desired parallels, namely, $AC||FG$. Applying the result of Problem 8b we also know $AC||EH$.

For $BD||GH$, one can argue in the following way: First, we have $\frac{f}{e} = \frac{g+h+i}{d+c+j}$ (an angle bisector of a triangle divides the opposite side in the ratio of the adjacent sides), and since the triangles PBA and PCD are similar we can conclude $\frac{g+h+i}{d+c+j} = \frac{i}{i}$. As we established above, $\frac{i}{i} = \frac{k}{l}$, and applying again that an angle bisector of a triangle divides the opposite side in the ratio of the adjacent sides, we get $\frac{k}{l} = \frac{c}{d}$. Therefore, we have shown $\frac{f}{e} = \frac{c}{d}$ and this proves $BD||GH$. Using a result of Problem 8b, which established that $EFGH$ is a rhombus, we also know $BD||EF$. Thus, we have shown that all the lines indicated in Problem 8 are parallel.

Problem 9: In triangle ABC, shown in Figure 6.34, where should point P in the interior of the triangle be placed so that it maximizes the product xyz of its (perpendicular) distances to the sides of the triangle?

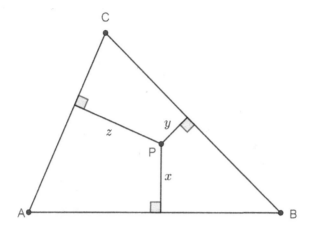

Figure 6.34

Problem 9a: Which point P on segment AB, shown in Figure 6.35, maximizes the product yz, which is the product of its perpendicular distances to the other two sides of the triangle?

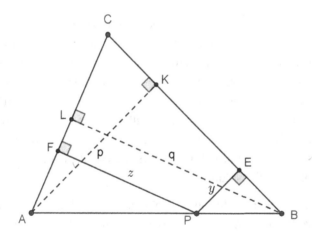

Figure 6.35

Let us denote $AB = c$ (constant), $AP = t$, $PB = c - t$, or expressed another way $0 \leq t \leq c$ is variable as can be seen in Figure 6.36.

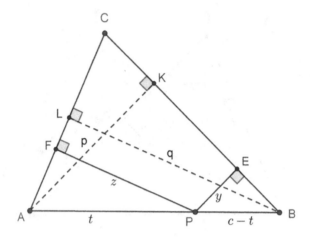

Figure 6.36

From the similarity of triangles APF and ABL, we have $\frac{z}{t} = \frac{q}{c} \Rightarrow$ $z = \frac{q}{c}\cdot t$ and for triangles PBE and ABK, we have $\frac{y}{c-t} = \frac{p}{c} \Rightarrow y = \frac{p}{c}\cdot(c-t)$. Hence, $y \cdot z = \underbrace{\frac{p}{c}\cdot\frac{q}{c}}_{\text{constant!}}\cdot t \cdot (c-t)$, and we want to know for which value of the variable t does this product become a maximum. Since the constant factor $\frac{p}{c}\cdot\frac{q}{c}$ does not affect which value of t yields the maximum of the product, as it only affects its own maximum value, we can omit this constant factor, and we have to maximize $t \cdot (c-t)$ with variable t. This is a very well-known situation: The maximum is achieved for $t = \frac{c}{2}$ (the square has maximum area of all rectangles with a given perimeter, formulated algebraically: The product of two non-negative numbers with constant sum c achieves its maximum, if both factors are equal, namely, $\frac{c}{2}$). Thus, the midpoint of AB is the solution.

Problem 9b: Use the solution of Problem 9a for solving Problem 9.

To solve Problem 9, we can apply the finding of Problem 9a to triangle $A'B'C$, where the line $A'B'$ is at a distance x above AB, as shown in Figure 6.37. Then the optimal position of P is at the midpoint of $A'B'$. That means (x again considered as variable) P must lie somewhere on the *median* m_c since the median for triangles ABC and $A'B'C$ is one line. Analogously, P must also lie on the *other medians*, thus, the solution of Problem 9 is the *centroid*, which is the intersection point of the three medians.

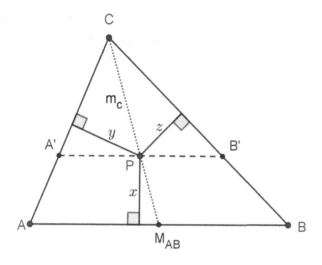

Figure 6.37

Problem 10: Prove that in a $2m \times 2n$ distorted chessboard, as we show it in Figure 6.38, the sum of the shaded areas equals the sum of the white areas. The meaning of "distorted" more precisely is as follows: On each side of the convex chessboard, we mark equidistant points ($2m$ or $2n$ equal segments), then these equidistant points of opposite sides are joined.

Figure 6.38

Problem 10a: Prove that in a "distorted" 2 × 2 chessboard, shown in Figure 6.39, the sum of the shaded areas equals the sum of the white areas. Here both pairs of opposite midpoints of the sides are joined.

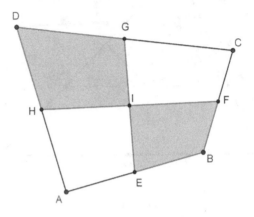

Figure 6.39

We will discuss two solutions:

- In the first, shown in Figure 6.40, the connections of the interior intersection point, I, to the four vertices are drawn, then one can see immediately that on every side, two triangles with equal area arise (they are not congruent): one white and one shaded. These triangles have equal bases and equal altitudes, such as $\triangle AEI$ is equal in area to $\triangle EBI$.

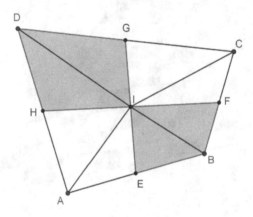

Figure 6.40

- For a second possibility, draw the parallelogram by joining the midpoints of the sides of the quadrilateral, as shown in Figure 6.41. Then one can see immediately that in the interior of this parallelogram, there is an area balance between shaded and white (in a parallelogram the diagonals bisect each other) and the remaining white (triangles AEH and FCG) and shaded triangles (triangles EBF and HGD) can be "pushed together," respectively, to a white and a shaded quadrilateral, which is in both cases similar to the original quadrilateral, thus, also these quadrilaterals have equal areas and in this case, they are congruent as well.

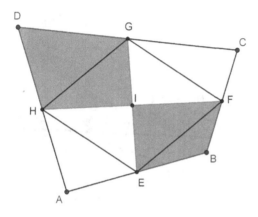

Figure 6.41

Problem 10b: Prove that in a "distorted" $2^m \times 2^n$ chessboard, the sum of the shaded areas equals the sum of the white areas, which can be seen in Figure 6.42.

First, we deal with the case where we have a 4×4 portion of the chessboard as we can see in Figure 6.42. Initially, the point labeled 1 can be fixed as the midpoint of HF and EG. Furthermore, E, F, G, and H are the midpoints of the sides. As the next step, the point 2 is the midpoint of $E1$ and PK, which is obtained by joining the midpoints of the sides of the quadrilateral $ABFH$. Analogously, the points 3, 4, 5 are also midpoints. Subsequently, we can conclude that the points 6, 7, 8, 9 are the midpoints of the corresponding line segments (diagonals of a parallelogram bisect each other). Altogether we see that all the interior intersection points are the points of quartering the corresponding joining line segments. With that knowledge, it is easy to argue for the equality of shaded and white areas in this case. In the quadrilateral $AE1H$, we have balance between shaded and white, as well as in the other three quadrilaterals ($EBF1$, $1FCG$, and $H1GD$). Hence, overall the same exists in the quadrilateral $ABCD$. But in order to conclude this, it is essential to know that the points 2, 3, 4, 5 are the respective the midpoints of the line segments joining the point 1 with the sides' midpoints E, F, G, and H. A similar way of arguing can be made in the case of an 8×8 chessboard and even more generally in the case of a $2^n \times 2^m$ chessboard.

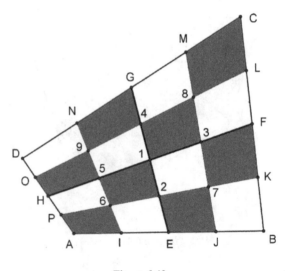

Figure 6.42

Problem 10c: Prove that in a "distorted" $m \times n$ chessboard resulting from any convex quadrilateral, all the interior intersection points divide all the corresponding joining lines equally, that is, the *equality* of the line segments on the sides transfers to the interior net of joining lines.

The corresponding "reduced" form of this problem is shown in Figure 6.43. If in a convex quadrilateral the opposite sides AB and DC are divided by the points S, and T in the same ratio $\frac{p}{q}$, and the opposite sides BC and AD by the points U and V in the same ratio $\frac{r}{s}$, then the intersection point P of VU and ST also divides these line segments in the ratios $\frac{p}{q}$ and $\frac{r}{s}$, respectively. In other words, the equal ratios on opposite sides transfer to the interior ratio in which the intersection point P divides the joining lines VU and ST.

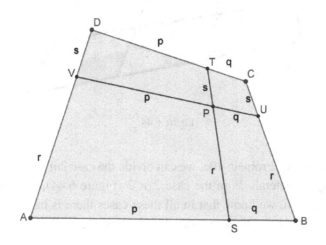

Figure 6.43

We can easily establish a "formula" for the points S, T, U, and V with respect to the points A, B, C, and D. For example, for S we get
$S = A + \frac{p}{p+q}(B - A) = \frac{q}{p+q}A + \frac{p}{p+q}B$.
Analogously, we get

$$T = \frac{q}{p+q}D + \frac{p}{p+q}C, \quad U = \frac{s}{r+s}B + \frac{r}{r+s}C,$$

$$V = \frac{s}{r+s}A + \frac{r}{r+s}D$$

And with that it is easy to verify that $\frac{q}{p+q}V + \frac{p}{p+q}U = \frac{s}{r+s}S + \frac{r}{r+s}T$ holds true, and this, in turn, yields the claimed proportions that the intersection point P of lines VU and ST divides these line segments in the ratios $\frac{p}{q}$ and $\frac{r}{s}$, respectively.

Problem 10d: Use Problems 10a, 10b, and 10c to solve Problem 10 (Figure 6.44).

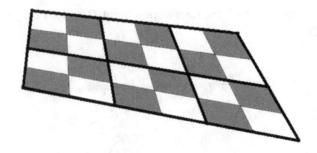

Figure 6.44

As a result of Problem 10c, we can divide the case $2m \times 2n$ into $m \times n$ smaller quadrilaterals as in the case 2×2 (Figure 6.44), and as a result of Problem 10a, we know that in all these cases there is balance between shaded and white, and hence, overall in the given quadrilateral.

Problem 11: Consider four equilateral triangles $\triangle A'B''C$, $\triangle A''B'C$, $\triangle AA'A''$, and $\triangle BB'B''$, as shown in Figure 6.45.

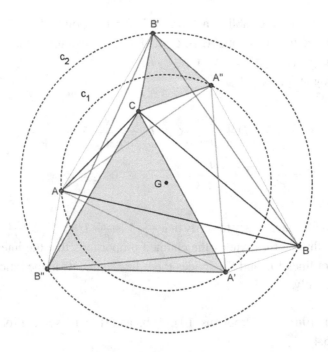

Figure 6.45

We begin by noting that point G is the centroid of triangle ABC. Then we construct two concentric circles c_1 and c_2 each with center at G, where circle c_1 contains point A, and circle c_2 contains point B. We then draw two equilateral triangles $AA'A''$ and $BB'B''$ with vertices on the respective circles, where A' denotes the point rotated around G by $120°$ counterclockwise, and analogously, B' denotes the point rotated around G by $120°$ counterclockwise. Prove that also the triangles $A''B'C$ and $A'B''C$ are equilateral, which we show in Figure 6.45.

Problem 11a: As we have stated above, $\triangle ABC$ has point G as its centroid. Now we construct two concentric circles, c_1 and c_2 with center G, one through A and the other through B, and two equilateral triangles $AA'A''$ and $BB'B''$ with vertices on the respective circles (here A' denotes the point rotated around G by $120°$ counterclockwise, and the same with point B'). Prove that $A'A''$ and BC bisect each other and that the same holds for $B'B''$ and AC, as we can see in Figure 6.45.

Extend AG beyond G by half of its length, then the new endpoint will determine the median for the equilateral triangles $AA'A''$ and ABC, which implies that this endpoint is the midpoint of $A'A''$ and also the midpoint of BC. Therefore, $A'A''$ and BC bisect each other. If we extend BG beyond G by half of its length, then we can see, analogously, that $B'B''$ and AC bisect each other.

Problem 11b: Use Problem 11a to prove Problem 11.
From Problem 11a, we can conclude that the quadrilateral $A'BA''C$ is a parallelogram, since the diagonals $A'A''$ and BC bisect each other. Analogously, we can show that quadrilateral $B''CB'A$ is also a parallelogram. Hence, $B'A = CB''$ and $BA'' = A'C$. But we also know by $120°$ rotations with center G that $B'A = B''A' = BA''$ and this means that $\triangle CB''A'$ is equilateral. And analogously, we can prove that $\triangle CA''B'$ is equilateral.

Conclusion

The principle of creating a series of preparatory problems is not always possible with every complex problem but, as we have seen, it is often useful. In some cases, it is not easy to give hints without divulging the solution.

On the one hand, good hints should not eliminate all the challenges of the problem; they can lead problem solvers to think creatively about the problem at hand. It is always desirable to experience mathematics as a process done by oneself. It is quite possible that this principle — looking for and creating a series of clever and elaborated preparatory problems — could be applied more often than we think. Problem solving in mathematics usually demonstrates the power and beauty of mathematics, in most cases a clever problem can be solved directly, yet in some cases, the stepwise procedures that we have demonstrated in this chapter provide a useful alternative. This could also be especially important for teachers presenting problem-solving strategies, as dividing a problem into smaller parts may significantly increase the chances for students to discover a clever solution.

Index

Printed in the United States
by Baker & Taylor Publisher Services